Collaborative Social Design with Mexican Indigenous Communities

This book builds on the work of anthropologists, designers, and ethnographers to develop an original methodology and framework for indigenous engagement and designer/non-designer collaboration in the field of social design.

Following a collaborative case study conducted over a five-year period between the author, project team, and indigenous artisans in Mexico, the book outlines the practical challenges of design research, including funding, logistics, relationships between designers and communities, failures, successes, and pivots. Social design literature has often focused on introducing important questions to the design research process, but fails to deeply interrogate and demonstrate how these theories inform research projects in action, which can then be open to misinterpretation, bias, and unintended harmful consequences. Centering the indigenous communities, this book provides a detailed and clear example of not just why, but how design and designers can work authentically and responsibly through different approaches and systems. The book examines the specific cultural, epistemological and socio-political history of Mexico as it relates to colonization and indigenous peoples, exploring the systemic influences of globalization and grounding the research in its unique context. It includes field notes, conversations with the indigenous artisan communities, workshops and prototypes to offer unique insight into a detailed, collaborative social design initiative.

This book intersects with the growing awareness of the necessity of decolonial approaches to design across the world and will be an important and useful study for academics, students and researchers in social design, sustainable development, cultural studies and anthropology.

Carmen Malvar holds a PhD in social design, is a member of the international GIS research group for Retail and Service Futures as part of the Design Research Society (DRS) and collaborator with the UNESCO Chair of Sustainability Barcelona with 20 years of experience on leading teamwork based on design practice. Carmen has worked on projects developed with indigenous communities in the highlands of México and is currently researching the social and ecological impact of some global industries and their backstage production, designing and implementing strategies where their production process and distribution chain shares the concern with stakeholders and the community. She is the founder of CADA Foundation, a platform for understanding cultural heritage through social design. Carmen works regularly as a consultant and speaker for various institutions, local governments, foundations, and academia.

Routledge Research in Social Design

Socially responsible design considers the broader environmental, economic, cultural, and political impacts of design on society. This series brings together cross-disciplinary research from a wide range of academics and design practitioners tracing the development of social design, its theoretical underpinnings and future trends. It welcomes contributions from established and early-career researchers examining social design through the lens of architecture, housing, product design and innovation, healthcare, education, cultural studies, sustainability, business, and design management.

Collaborative Social Design with Mexican Indigenous Communities
Critical Craft and Transformative Practices
Carmen Malvar

For further information about this series, please visit: www.routledge.com/Routledge-Research-in-Social-Design/book-series/RRSOCDES

Collaborative Social Design with Mexican Indigenous Communities

Critical Craft and Transformative Practices

Carmen Malvar

Routledge
Taylor & Francis Group
LONDON AND NEW YORK

earthscan
from Routledge

First published 2023
by Routledge
4 Park Square, Milton Park, Abingdon, Oxon OX14 4RN

and by Routledge
605 Third Avenue, New York, NY 10158

Routledge is an imprint of the Taylor & Francis Group, an informa business

British Library Cataloguing-in-Publication Data
A catalogue record for this book is available from the British Library

Library of Congress Cataloging-in-Publication Data
Names: Malvar, Carmen, 1967– author. | Castillo Cisneros,
María del Carmen, writer of foreword.
Title: Collaborative social design with Mexican indigenous communities :
critical craft and transformative practices / Carmen Malvar.
Description: Abingdon, Oxon : Routledge, 2023. |
Series: Routledge research on social design |
Includes bibliographical references and index.
Identifiers: LCCN 2022045852 (print) | LCCN 2022045853 (ebook) |
ISBN 9781032420134 (hardback) | ISBN 9781032420172 (paperback) |
ISBN 9781003360803 (ebook)
Subjects: LCSH: Design–Social aspects. | Indian handicraft industries–Mexico. |
Indians of Mexico–Economic conditions.
Classification: LCC F1219.3.I5 M35 2023 (print) | LCC F1219.3.I5 (ebook) |
DDC 338.4/77455089972–dc23/eng/20221014
LC record available at https://lccn.loc.gov/2022045852
LC ebook record available at https://lccn.loc.gov/2022045853

ISBN: 978-1-032-42013-4 (hbk)
ISBN: 978-1-032-42017-2 (pbk)
ISBN: 978-1-003-36080-3 (ebk)

DOI: 10.4324/9781003360803

Typeset in Bembo
by Newgen Publishing UK

To my family, and in special mention to Alba, my daughter, for her encouragement. And to all the people that have made this project possible for others to continue the road ahead.

Contents

List of Figures ix
Foreword: Collectivizing Design by María del Carmen Castillo Cisneros xii
Foreword: The Other by Maddalena Forcella xvi

PART 1
The Framework 1

1 What Does It Mean to Collaborate? A Study in Power 3

2 Framing the Practice: Globalization, Social Design, and Proximities of Design 22

3 Articulating Cultural Differences: Indigenism, Politics, and Social Context in México 43

PART 2
The Journey 71

4 Design as a Practice of Correspondence: El Hacer—The Making 73

5 Field Notes: Interviews, Workshops, and Prototypes 113

Closing Chapter: Exploring Continuity, Measuring Impact, and the Legacy and Future of Social Design Practice 177

Afterword: Proximity and the Ethics of Engagement by Barbara Adams 190
Index 195

Figures

1.1	Family walking through the Oaxaca Mixtec Region	4
2.1	Infographic: Communities' Permeability and Resilience	28
2.2	Making together at workshop	30
2.3	Relationships through and within the space. Left Santa Maria Yavesía. Right: Santo Domingo Tonaltepec	33
2.4	Left: Photo of artisan Juanita Bautista explaining the traditional technique of pepenado. San Pablo Tijaltepec. Right: Artisan Gregoria Cruz explaining the process of firing clay pieces. Santo Domingo Tonaltepec	34
2.5	Greetings and sharing food [Digital video]	35
3.1	Mixtec textile samples	63
4.1	Infographic Community's core; Ethos, ethics and aesthetic designed by Rut Martín and Carmen Malvar	76
4.2	Visiting communities—questions of power, gender, and local authorities	79
4.3	Team layout. Basic infography	83
4.4	A team meeting at the café center in Oaxaca	87
4.5	A permanent inaccessible road section	89
4.6	Artisans of San Pablo Tijaltepec observing chosen images	97
4.7	San Pablo Tijaltepec, artisan Rosalía Bautista. Cloth dolls made during workshops	98
4.8	Organizational routines during design workshops: Artisans Rosalía Bautista and Rosario Nicolás being observed by the oldest member (grandmother) sitting in a routine circle dynamic for activities during design	99
4.9	Experimenting together with new materials artisans had never used before	100
4.10	Working environments: open public and private spaces. Initiation of workshops	102
4.11	Some samples of the resulting prototypes	106
4.12	Stills from excerpts, picturing sound and movement through the material	107
5.1	Notes on field trips and meetings	114

5.2 José Martínez spinning and weaving to obtain Creole Silk 118
5.3 Map tracing travel conducted during this Ethnographic Research phase 124
5.4 Interior spaces at CECATI 133 in San Cristóbal de las Casas, depicting metalsmithing artisans at work 125
5.5 Fine examples of wrought iron, zoomorphic door handles abstractions of a bull and an armadillo 128
5.6 Detailed craftsmanship in an edge with openwork and a pendant, examples of jewelry and the production of fine decorative objects 129
5.7 Still workshop at CECATI 166 130
5.8 The circular dynamics of chairs and assembly with regard to space 133
5.9 Woman holding her drawing at first oral workshop 137
5.10 Diagram by Carmen Malvar showing off-center position in relation to communities in our approach 140
5.11 Children clay sample 148
5.12 Nazaria collecting the image from her work from previous workshop 150
5.13 Carmen and Gregoria Cruz reading through the artifact dossier during the trip back to Gregoria's home 150
5.14 Sample from Sebastiana García 151
5.15 Experimenting with new materials 154
5.16 Working with 3D/play dough on face expressions 156
5.17 Artisan Nazaria and facilitator Diego, revising utilitarian and decorative commercial pottery at workshops in Santo Domingo Tonaltepec 161
5.18 The house of Rosalía Bautista prepared to hold workshops 162
5.19 Rufina Ruíz and the exchange of tools and techniques with artisans at the workshop 163
5.20 Natividad and Sebastiana García taking patterns out of shapes 163
5.21 Natividad and Sebastiana García reusing their old sewing machine arranged by Rufina Ruíz 164
5.22 Cuts and shapes developed at workshops with women of San Pablo Tijaltepec 165
5.23 Laura Quiroz and Maddalena Forcella reviewing new samples commissioned and brought by Laura from San Pablo Tijaltepec 166
5.24 Artisan Demetria Cruz working on the production of pieces 167
5.25 Firing. A group gather and communal activity 167
5.26 Result of some early samples already burned and finished 168
5.27 Diego reviewing fired pieces commissioned 169
5.28 Interior of a section of the temporary office of CADA 169
5.29 Working together. Exchange of tools and techniques with artisans 170

5.30 Some samples of resulting prototypes from San Pablo Tijaltepec with the use of sewing machine for finishing 170
5.31 Artisan Sebastiana García. Member of the collective Ñaa Ñanga 172
5.32 Women in the 2018 workshop observing a textile from El Camino de los Altos at the new construction of Rosalía's house 173

Foreword

Collectivizing Design

María del Carmen Castillo Cisneros

Carmen Malvar, who is quite aware how this subject is an exceptional source of irritation for me, for both personal and anthropological reasons, must be extremely brave to invite me to write a foreword to her book. Deep into the 21st century, we cannot write off the controversies surrounding design and indigenous peoples that are at the heart of the polemic that has been set in motion by capitalism. My long-standing ethnographic work in the state of Oaxaca has allowed me to see in real time the revitalization that craftsmanship has had in terms of materiality, creativity, and skill as it emerges as a highly fetishized resource in the current global market. The impact, not only economic but social and cultural as well, must be analyzed and scrutinized with a designated magnifying glass for questions that have to do with national, identity, and mercantile discourses that should not be eschewed from the configuration of new narratives.

And this is why *Collaborative Social Design with Mexican Indigenous Communities* represents a work that not only captures the experience of more than five years of work with two Oaxacan communities, namely, San Pablo Tijaltepec and Santo Domingo Tonaltepec, but also presents us with an opportunity to question practices that have emerged worldwide as a byproduct of the "ethnic boom" with the necessary subsequent reflection that this entails. Although the author doesn't offer a specific formula, she dares to link design and craftsmanship by exploring the context of production and rethinking patterns that pertain to the creative process.

Everyone knows that in an increasingly homogenized and individualistic world, the act of collective making while drawing on memories shared by the community is misconstrued and undervalued. But life in other latitudes shows us that there are still communal objectives of inhabiting space through dialogues that resist the forces of a modernity that is bent on turning them into monologues. As Raúl Zibechi (2022)[1] rightly argues,

> the autonomies walk in the most diverse and remote corners of Latin America. Among native peoples, blacks and mestizos, in the fields and in cities, between peasants and workers, hand in hand with women and elders who push them from the most elemental need to preserve life, to

continue being communities. They each walk at their own pace, with their own rhythm, moving their feet to the beat of their dreams or their most urgent needs.

So, what is the task that falls to designers who work in these communities? I would say first and foremost, a fundamental respect for other ways of being in and inhabiting the world and relating to cultural artifacts that pertain to the community, determine the survival of the community and are the product of a collective creative action that extends and bolsters the consciousness of a given community's existence. In this sense, Malvar maintains that her work tries to present a more complete understanding of the artisanal process, its artifacts and what they represent for their respective communities, culture, and way of life.

Let us not forget that above all, it is always about people at the end and cultural artifacts have always functioned as connectors, objects that bond human beings through time and space in such a way that, as is well stated in this work, it falls to the communities to decide the best way of integrating and adopting objects in function of their worldview and acknowledgment of personal and communal objectives. As I have expressed on other occasions, we are not talking just about textiles, ceramics, or handcrafted designs that exist in a sociocultural vacuum. We are also talking about the co-existence of communities with the current challenges facing indigenous territories: conflicts over water sources, forests, jungles, and habitat where specific flora and fauna are in danger and in dispute. We are, furthermore, talking about the seizure of communal lands, caves, and other sites associated with supernatural entities that are meaningful and socially significant to the community, snatched up for megaprojects. Clearly, this is a discussion about many things and more, and it must be emphasized that crafts are present in all this as much more than just a particular style of embroidery or woodworking or design. Their intangibility belongs to a specific way of thinking, structuring, and understanding a changing world that is not only unknown most of the time but stands in contradistinction to hegemonic Western models perceived as unique.

But what are we talking about when we talk about social design? Although the author discusses social design as a human process that helps us understand and interact with the world and among ourselves, I wonder if in that sense not all design is, by definition, social. Thinking out loud, as I make my way through this work, I ask myself about the usefulness of semantic categories and the consequences of attaching a nomenclature that may only serve to mark an apparent differentiation that shows how an unnuanced approach can often lack friendliness toward different existing cultures. Thus, this book offers us a lot of material to question this so-called social design and, above all, to reflect on what the author herself calls proximities of design in working with communities.

By way of these proximities, Carmen Malvar calls attention to the time-space, manufacturing practices, choice of materials and the social context wherein indigenous communities exist as a starting point for an approach that, above all else, considers artifacts as non-traditional traditions, as open-ended living cultural expressions that transform and adapt. In this vein, the author underlines the living nature of things, how they inhabit social environments while establishing relationships with the subjects of their culture. For this reason, artifacts, things that are made by hand, whatever is "alive" is also a receptacle of memory, a story that the author highlights by taking a historical tour of Mexican indigenism to illustrate how craftsmanship in our country plays a part in a shared national identity. This work presents a detailed outline of the social and political factors that have shaped indigenous narratives and politics in México; it analyzes how indigenous cultures and their materialities have been used as part of a nationalist narrative. This work also addresses, on numerous occasions, the relevance of oral tradition to craft creation, memory, identity, and the survival of worldviews. Here, oral tradition is linked with the dynamics of resistance and resilience that are reflected in the objects that a society conceives as a fundamental part of its culture, and this is a source from which design has much to learn.

However, the methodology exposed throughout this work reveals the importance of collaboration. The author notes that the time allocated to projects of this nature must be generous and not subject to short periods that can only allow for a superficial understanding of the socio-cultural reality of communities. This adherence strikes me as extremely significant when approaching the heterogeneity of cultural manifestations that underlie the different ethnic configurations of our country.

On another front, I cannot fail to mention a concern that arose throughout the author's research process and also marked a milestone in her own approach to design. Namely, the assault by international companies, the imitations of traditional textiles as well as other misuses and exploitations, and the community response and other alarms that were set off in Oaxacan localities in the wake of the plagiarism of the Tlahuitoltepec blouse in 2015. For Carmen Malvar, these events created an unease and raised questions about artisan ownership of their own work as well as an urgency to communicate to artisans an accurate representation of what was occurring on a global sphere in terms of the risks they ran when showing their crafts to a larger public. After the author brought up these issues to local authorities in San Pablo Tijaltepec, they requested an anthropological report on the appropriation of their blouse from the INAH authorities, an institution where I work and was therefore given the opportunity to attend the communal hearing.

As I finished reading the book you are holding in your hands, the case of the blouse came to reflect a meeting of the minds between myself and the author. We both opt to link ourselves to research processes to which we have been invited by the community. It seems obvious to say, but it does not always happen this way. My collaboration with the anthropological opinion of the Tijaltepec blouse was carried out at the request of authorities from a community faced with the uncertainty of design appropriation and hoping to obtain a document to secure artistic and cultural ownership of the blouse, its pepenado technique and design features. The opinion was written as part of the Textile Memory

Project that at that time was coordinated at the INAH-Oaxaca Center with the collaboration of Amapola Rangel Flores, a student majoring in anthropology who later would write her thesis on the Blouse.

It has been a privilege and a challenge for me to take on this foreword that not only provokes a self-critical reflection and observation of design and its practices, but also makes me consider the principles and challenges of my own discipline; anthropology, full of accents and social tensions that seek to rethink, adapt and take new approaches into account. From this experience with Tijaltepec, as a social scientist, I declare myself in favor of an on-demand anthropology, or what Valladares (2021)[2] has called a "community anthropology," a type of anthropology that is in tune with collaborative models and epistemologies that postulate the establishment of horizontal dialogues with interlocutors with whom we are linked in the research processes that we undertake. Being invited as anthropologists, or in this case, as community designers could become another way of carrying out our professions in and with communities who are exercising their autonomy. An anthropology, or a design, that follows these guidelines requires new dialogues and commitments in terms of solidarity, and the ethical and academically committed collaboration with the counter-hegemonic social processes that build democracies and pluralities.

As Malvar puts it, *Collaborative Social Design with Mexican Indigenous Communities* provides a useful case study for design practitioners, educators, and researchers to break new ground and form a crossroads between fields that are invested in collaborative work, design practices, and material culture.

This book proposes an approach and tool that give future generations a base on which to create new design methodologies in accordance with host communities. Collectivizing design, as this work suggests, offers a way of respecting the cultural plurality of a country that cries out for other possible futures for indigenous peoples. Much remains to be done, Carmen Malvar has pointed out the path that will make us part of the process and shown us the necessity of walking together as community.

Notes

1 Las autonomías indígenas se están convirtiendo en alternativas de vida y en referencias políticas. Raúl Zubechi, 2022. https://agenciatierraviva.com.ar/las-autonomias-indigenas-se-estan-convirtiendo-en-alternativas-de-vida-y-en-referencias-politicas/.

2 La antropología comunitaria. Una nueva relación de investigación en y con los pueblos indígenas. Laura Valladares. Alteridades 2021, 31 (62):13–24. https://alteridades.izt.uam.mx/index.php/Alte/article/download/1279/1340.

Foreword

The Other

Maddalena Forcella

The first time I spoke with Carmen Malvar was in a short virtual chat in which she discussed her research project and told me about her plan to travel to Oaxaca in the near future. Perhaps something she said, or the way in which she talked about her project, persuaded me to introduce her to a number of friends and colleagues who I thought would be interested in collaborating in a project that struck me as promising and, to my mind, very novel. I was especially interested in the development of a theoretical reflection on fieldwork with artisans and indigenous communities—something I had practiced empirically for many years as well as the possibility of comparing my experience with that of friends and colleagues with similar experiences. The author's interest in knowing the background of the collaborative practices of artisanal design in México helped stir up some serious reflection, and eventually resulted in additional trips and meetings with the protagonists of pioneering initiatives in Santa Clara del Cobre and Chiapas. Both are documented in this book.

There is no doubt that the experience was enriching; many of the after-dinner talks during which we developed a profound mutual understanding as we talked about visions, work ethics, programs, and dreams, are recorded and compiled in this book. Reading them again today fills me with emotion, and I believe they will be a valuable testimony for anyone interested in a subject that has become increasingly current. In recent years, the indigenous and artisanal communities of México have been subjected to increased interventions by external agents who, for the most part, are oblivious to the cultural content of artisanal artifacts which are instead reduced to the status of commercial objects and valued only for the profit their sales can generate, without a second thought for the ethics of their production, or their creators, or the original culture that produced them.

I have personally had the good fortune, for many years now, to visit many communities in México on behalf of various institutions and organizations, sharing experiences with weavers and artisans, working on the development of new products focused on rescuing materials, techniques, and traditional iconographic motifs.

At first it was not easy. As a product of a European, urban culture, it took me time to adapt to the unfamiliar environment. It took years of coexistence and reflection to—I think I can now safely confess—relate without prejudice and horizontally with people who on the face of it appeared to be so different.

One experience, in particular, has remained emblazoned in my memory as a revelation that has marked my professional life and helped me understand in an intimate way what is meant by cultural difference. This experience brought me to the realization that it is possible to feel intimacy and share moments of rare intensity with people one recognizes as different. This happened in front of the Pacific Ocean, in the company of the daughters and granddaughters of the expert Ikoot weaver Justina Oviedo. It happened at sunset in 2005 in San Mateo del Mar, Oaxaca. After a long ramble through dunes and palm trees, we sat down on the sand and watched the setting sun, and perhaps inspired by the type of emotion that often surfaces when looking at the vast expanse of the sea, we began to talk about the creation of the world, the universe and religion. The dialogue was quite surreal: I held an atheist and rationalist point of view, claiming that life and everything else had come from the sea, while Cristina, the eldest of Justina's daughters, tried to convince me that of course not! God had created everything! The discussion lasted a long time, the sun had almost disappeared, and then suddenly, Cristina and her sisters decided they wanted to give me a blessing to protect me on my return trip, a ceremony. I don't remember it in detail—I think we all held hands; but what I do remember is the feeling of deep spirituality that reconciled us all in the end. The moment was truly extraordinary and unique.

Carmen points out that the concept of social design should be understood as an action of "doing together," with the aim of understanding local identity and exploring the ways in which the practice of design can generate changes that will benefit artisan communities in relation to global markets. The exchange of knowledge and values between communities and design professionals must occur in a context of equality. But this context of equality is, in my opinion, something that very rarely occurs spontaneously between actors coming from different cultural systems. I, therefore, ask you to allow me to briefly summarize the concepts raised by the great scholar Tvetan Todorov[1] when analyzing what he calls "the question of the Other."

Todorov tells us that there are at least three axes around which the problem of alterity is constructed, namely:

CONQUEST: in the first place there is a value judgment (axiological plane): the other is good or bad, I like them or I don't like them. The value judgment is automatically understood as a function of superiority.

LOVE: second, there is the action of approaching toward or moving away from the other (praxeological plane) refers to action. Either I embrace the values of the other, identify with them, or assimilate the other to me and impose my image on them, as exemplified by the indigenous policy of México until the 80s of the 20th century. There is also the instance of a third path: neutrality or indifference.

KNOWLEDGE: third, I know or ignore the identity of the other (epistemological plane) in which there is an infinite gradation between nominal cognitive states and advanced cognitive states.

Neither conquest nor love can lay the foundation for a true recognition of the other as both these pertain to the plane of egocentrism wherein one identifies one's own values with values in general and the "I" is conflated with the universe instead of being situated from a "perspectival" viewpoint in which each may be in accordance with their own values and not judged according to a single, monolithic ideal. So in lieu of taking and destroying, or assimilating, Todorov tells us there is the way of understanding. There is a positive or neutral form of exchange with the other, and it is communication. It is possible to establish an ethical criterion on the basis of which an opinion shaped by experience and influence is expressed: what is essential, the author tells us, is the awareness of which influences are imposed or proposed. A thing is not imposed when the possibility exists of consciously choosing another thing.

Keeping these concepts in mind and trying to maintain an honest awareness of the position from which we relate to the other is a valuable tool and guide to respectful and responsible action in design practice as well as in life in general. I know this was the author's attitude from the moment she arrived in México, an attitude she maintained throughout the years it took to complete this investigation. And now that I think of it, this attitude was what captivated me from the start and enabled us to lay the foundations of what would become a long professional and personal friendship.

July 2022

Note

1 Tvetan Todorov "La Conquista dell' America - il problema dell'altro" Einaudi.

Part 1

The Framework

1 What Does It Mean to Collaborate?

A Study in Power

In México, indigenous cultures have been scrutinized through the lens of a national integrationist policy since the end of the Revolution. A number of decisions have been made at the highest levels of Mexican government and society which have led to a national implementation of initiatives aimed toward homogenizing the beliefs and practices of distinct cultural groups throughout the country. Many of these social, political, and cultural decisions involve pressures to force changes to arts and artisans' heritage based on an approximation of a unified Western standard.

With many communities facing the threat of a forced cultural, social, and economic integration into a more homogenous global network, it is important that we contextualize the cultural and historical narrative in México based on the cosmovision of its original inhabitants. The complex story of México is one of conquests, revolutions, and a heterogeneous contemporary society that faces pressure toward homogenization today more than ever as politicians and other forces of power make decisions to conform in an age of global production. Today, México remains a country that reflects many sharp contrasts. The unequal access to public assistance or opportunities faced by a culturally and ethnically diverse Mexican population may be most clearly defined. In many small Mexican cities and towns, a traditional and patriarchal structure governs daily life based on balances of social and political power derived from gendered norms and assigned roles. It is crucial that one recognizes a variety of cultural, social, economic, and circumstantial elements that intersect and diverge across local contexts in México in order to reflect on the possibility of fostering resilient communities in our age of globalism.

México's Oaxaca region represents one of the most culturally diverse regions in the world. The indigenous artisan communities that reside and work in Oaxaca have a deep connection to place, are geographically isolated, have limited access to resources, and have been severely marginalized by the mainstream vision (see Figure 1.1). There are 17 ethnic groups living in the state of Oaxaca itself (93,000 km^2), where 137 of the 364 known languages in México are spoken. The majority of groups living there call themselves either "the people of the clouds" or "the people of the rain." Zapotecs, Mixtecs, and their neighbors maintain traditions of African and European influence (De Avila,

DOI: 10.4324/9781003360803-2

Figure 1.1 Family walking through the Oaxaca Mixtec Region. An example of geographical isolation. CADA Foundation®, 2014.

2013). The territory is also physically distinct with a mountainous region of great geographical complexity in the center of Mesoamerica.

Between 2013 and 2018, I embarked to develop several workshops in different local indigenous communities throughout the Oaxaca region. These communities had requested support for gaining greater market exposure through local organizations working within the region. My aim in beginning this project was to provide the requested support while avoiding the potential loss of local identities and to foster the reinforcement of their own communities through the use of social design practices. This process involved sharing tools that allowed for the creation and production of artifacts that could then be integrated into a globalized market as long as the artisans were mutually interested in doing so. In the contexts of the local and the growing global economy, these workshops were oriented around collaborating with people in places that seemed to be a-capitalist cultures sometimes on the verge of disappearing. In these contexts, our workshops relied on relationships of trust, the mutual exchange of knowledge, and the recognition of authorship and integrity of collaboratively created work. It is important to clarify that this project was not based on altering any physical artifact, but rather to reflect on a communal practice of social design as a potential tool for building resilient local identities in the age of globalism.

In my perception of applied design, I often considered the effect of economy on culture and the sorts of disbalances provoked by forces of globalism. However,

I also considered how the global economy might serve as a powerful tool for artisan communities. I began by asking myself how a collaborative design practice might facilitate greater access to resources for these communities interested in engaging with the global marketplace on more equal terms and by expressing their cultures on the world stage.

In order to envision the global economy as a tool for creating resilient communities, myself and a team of design practitioners—serving as project facilitators—considered how the de-colonialist framework attempts to open the door for non-western perspectives of design. Our approach to the workshops was based on ideas of co-creation and a mutual exchange of information between all participants rather than on ideas of activism or "providing help." We facilitators and artisans made decisions during workshops and meetings in a democratic, communal, and qualitative way. We facilitators endeavored to generate a dialogue between different forms of design practices with and within communities through the creation of situations for artifact exploration. Ultimately, these situations led to the artisans' consideration and eventual transformation of their artifacts to represent their cultures, skills, and values for a wider market. We facilitators believe that such a result was only possible by generating a respectful correlation, or a dialogue between multiple parties, by a willful exchange and not through imposition.

In this chapter, I outline the objectives and foundational theoretical concepts that guided the project and which led to this book. I also tell the story of how I developed a core design team who facilitated collaboration in the workshops. This may serve as useful information for those looking to build international research teams and poses key questions for those wanting to do social design work with indigenous communities. I also explore some of the key power dynamics that I faced while working with artisan communities in Oaxaca, México.

Project Development

The development and implementation of community workshops discussed in this book aimed to facilitate collaboration between artisans and design practitioners. The practice took shape over several stages between 2013 and 2018. Based on communities' expressed interest in gaining market exposure, a team of designers and myself selected communities through ethnographic research that included exploratory visits and return visits to conduct single-day workshops. A second phase of the project began as a collaborative and an evolving design process concerned with developing scenarios for long-term engagement with these communities. The third phase focused on means of prototyping artisanal production and initiating market exposure. The resulting workshops during this process were held and developed with artisans to collaborate with materials and knowledge to produce samples, refine finished goods, and prepare objects to be marketed in a wider global context. We facilitators documented our collaborative work throughout all three phases of the project only once we informed each community of the project's purpose and we were granted their consent.

Between 2013 and 2014, I traveled to Oaxaca to conduct an ethnographic exploration to map the Oaxaca territory both geographically and contextually. Exploratory visits during this time included meetings, interviews, and initial oral workshops within rural Oaxacan communities. I also traveled to San Cristóbal de Las Casas in Chiapas, Mexico Distrito Federal, and Santa Clara del Cobre in Michoacán. These events connected artists, artisans, thinkers, designers, and anthropologists to kindle ideas, generate situations, and discover synergies. During this period, I gained great insight into local Mexican cultures, as well as social and political issues. I made my first contact in Oaxaca during this time through an alumnus who had invited me to visit her native Oaxaca. Fátima Díaz, an architect and designer, introduced me to her mother, a senior DIF[1] official, who facilitated transportation logistics and contacts to local authorities during some of the first workshops. During this time, I also met a well-known textile artist named Maddalena Forcella who dedicated her life work to collaborate with indigenous communities in México.

As I conducted this exploration, I also gathered a group of experienced design practitioners through personal connections and based on their previous involvements in projects and collaborations with local artisans and communities. All of these practitioners would serve as workshop facilitators due to their mutual interest in design as a practical action and their years of experience working with diverse communities. The role of this team, as facilitators, would be to participate in the workshops alongside artisans to link their experiences in the academic field of design with practical processes of design in order to foster collaborative spaces.

The team and I endeavored to conduct these workshops with ideas of social design as a core of our practice while also considering historical and cultural contexts. This practice aimed to understand local identities and explore ways in which social design could generate changes in socio-economic and political realities faced by local artisan communities. During the initial exploratory phase of this project, I found that one's mastery of craftmanship is well regarded and socially recognized in México. Individuals who are considered "*maestros artesanos*" possess deep cultural knowledge, respected social positions, and are understood to be conveyors of ancestral knowledge within the community. As researchers and design practitioners, we thus recognized the importance of including these artisans in the dialogue of a design process. This means that we considered artisans' design practices, products, and ideas as their own and for their own means. It is important to note that the objectives and outcomes of the workshops were not determined by myself or by the other facilitators, but by the artisans and their respective communities.

The team discussions in early meetings opened space for a range of collective critical engagement and allowed us all to express our fears and excitements as well as our tensions and disagreements. Based on our previous practical experiences, we agreed that it would be important to consider the situations we encountered more comprehensively by exploring deeper details of spaces, bodies, and emotions—or, through what I call "proximities of design." This also involved our discussion of the ways in which our experiences of knowing and doing in the field of design practice afforded our positions within the project a maneuverability that artisans may not have been exposed to previously.

Throughout the book, I refer to "proximities of design" as situations when we design practitioners could understand and frame events between people, places, and things within a context of design. Clifford Geertz (1983) coined the term "approximation" as a method of reading or translating situations. Making an approximation involves thinking about the ways in which people communicate with each other and act in an environment through their gestures, movements, and practices. I conceive of an approximation as an agreement pertaining to how a practice of correspondence can occur between people and the physical world. In the context of globalism, pressures to change cultural traditions for newer or "more efficient" ways of doing things can cause unequal terms of participation depending on location and identity. In response, this project defines a practice of correspondence as a set of conditions that allow dialogue without fear or distrust and that permit an exchange of knowledge between all participants. Proximities of design, then, refer to how events are framed through collaborative discussion and "the making"[2] that happens in a given space as it is occupied and transformed by a variety of individuals who belong to a broader community. As a tool for framing events in this project, "proximities of design" helps to explain the events that occur in an environment for articulating identity through design—whether it be in terms of a community or an individual. This is a collaborative process, and thus, relationships or expressions of exchange serve as the most powerful tools.

This project was based on the concept of design as a practice of correspondence, or reciprocity. Design wields an incredible power to transform society as an act of service. As Susan Yelavich posits in her book *Design as Future-Making*, acts of design and their consequences are, at their core, political acts (Yelavich and Adams, 2014). Adams widens this scope of discussion about participation through correspondence by recognizing the practice as including the spiritual and cultural values that comprise a community's worldview. During the course this project, I wanted to encourage the emergence of new sorts of social interactions through what I call acts of "the making." We workshop facilitators operated under the belief that knowledge of "the making" process can embody emotional connections and contribute to enriched cultural discussions that empower communities and preserve their legacies. We facilitators relied on a practice of design through "the making" as a method for interpreting and "approximating" a social group's values in order to collaboratively develop feasible outcomes aligning with the artisans' goals. This structure of social organization is a system of interactions among people and between people, places, and their artifacts. We facilitators believe that it is the complexity of these interactions and relationships that tend to guide behavior and decision-making processes.

It is also important to note that the project relied on a "practice-based" approach. Practice-based research has previously been defined as "an original investigation undertaken in order to gain new knowledge partly by means of practice and the outcomes of that practice" (Candy, 2006). Embracing this formulation of a practice-based methodology, we facilitators focused on the process of doing and the outcomes of that process. Through our observations of cultural and political power relations surrounding the creation of artifacts in local Mexican communities, this work attempts to present a more complete understanding of artisans' creative processes, their artifacts, and what they represent for their respective communities, their culture, and their way of life.

Project Phases and Facilitator Team Development

Whereas it was once unusual for designers to share and adapt work methods from other fields, that practice is now very relevant to conducting research that encompasses design and approaches social problems. In his work *Prototyping the Social*, James Hunt (2011) compares ethnographic practice with the designer's profession regarding their respective responsibilities to society. While ethnography concerns itself with studying the present in detail, it does not typically include intervention. Its aim is rather the production of descriptions, which are by nature retrospective. Bronislaw Malinowski's model for ethnography was oriented around observation, and aims to explore how beliefs, behavior, material objects, and the constitution of social groups can also be seen from the perspective of their interrelationship (Malinowski & De La Fuente, 1957). Donald Schon's (1983) concept of "reflexive practice" can be used to bridge the gap between the distinct concepts of making and cogitation (Schon, 1983). Schon proposed the need to generate an intuitive, singular process able to provoke reflection in conflictive, complex, and uncertain situations. Within the context of the present work, I define my position as a reflexive practitioner along these lines because such a position allowed me to delimit the scope, format, and direction of the project's workshops based on my observations during their development and of the practices applied within them.

As part of the ethnographical phase in February 2013, I visited five towns in the vicinity of Oaxaca de Juárez, the Province´s Capital. The aim of these visits was to gain a peripheral view into those specific artisan communities who had expressed their interest in gaining exposure to new global markets. Local crafts there involved the use of materials such as clay, wood, and embroider silk or palm fibers.

The first town we visited was San Marcos Tlapazola, well known for its particular terracotta-style red clay pottery. Traditionally, San Marcos has long been a supplier of pottery to the Central Valleys and the Tlacolula market. Although there are about 300 potters in the village, they work sporadically. However, since the late 1980s these potters have been experimenting with new forms, which has led to better sales and recognition for the village. Artisans market their wares on market days at Tlacolula and most recently also at the crafts market at Oaxaca de Juárez.

Another town that we visited which was working with clay is San Bartolo Coyotepec. Local pottery there is unique, not only because of its black color, but also because of its texture. Artifacts are so fine and well burned that they become useless for cooking. However, for centuries the pottery of Coyotepec has been recognized for its capacity to containing liquids. Barro negro, as their ware is known, also has a very particular finishing process: when pieces come out of the pit (as they are not fired in a kiln), polished pieces are glossy black and the unpolished ones have a matt grey finish. This process was developed after 1930s when the town's economy was affected by sales of aluminum utensils, well pumps and both aluminum and plastic wares. Forced to innovate, potters began to burnish their pieces by rubbing stones against artifacts, which gave rise to the polished black clay that has made them so famous. Nearly 95% of Coyotepec's population is involved in barro negro production. Thanks to their robust sales, most of the potters live comfortably from their trade. Countless shops in Coyotepec sell local pottery and it is also known across the Country. Other towns in the region are well known for their wooden crafts, whether they be decorative or functional items. San Martin Tilcajete's artisans carve wood Alebrijes—mythological and zoomorphic characters—out of softwood. These carvings are mostly polychromed and are well known in other parts of México. Initially, Alebrijes made in the region around Mexico City were wooden, while those made in Oaxaca were first constructed out of paper and then were later made of wood for durability. In Oaxaca, the tradition of carving wooden animals is very old and linked to Zapotec cultural heritage. In this cultural cosmovision, the carvings represent spiritual guides—nahuales—in the life of each person according to the day of their birth.

During the exploratory visits mentioned earlier, the team and I selected communities based on our assessments of craftmanship quality, exposure to outside influences, desire to market access, and the risk of disappearing techniques. Most importantly, though, we also selected artisan communities where designers acting as facilitators had previously established collaborations with artisans or where designers maintained cultural and linguistic relationships. These familiarities allowed the team and I to more easily gain trust through novel methods as well as build upon existing relationships where no collaboration was previously conducted. For example, we selected San Pablo Tijaltepec because the community sought the support we could offer, and because of their unique handcrafted embroidery called "pepenado." Also, Laura Quiroz, a member of our facilitator team, had personal connections with community members there.

San Pablo Tijaltepec is a small village laid amidst the dry and rugged landscape of *La Mixteca* (in local language, mixteco Ñuu Savi). Shared by the states of Puebla, Guerrero y Oaxaca, the region is culturally distinct with a separate political and economic organization. In spite of—or maybe because of—its isolation, San Pablo Tijaltepec is unique for its embroidered cotton textile technique of Pepenado or Penado: a pick-up weave with gathered pleats which is then embroidered with polychromed figures and symbols of topography and nature. The community is extremely difficult to access because its road system is not maintained. Here, local inhabitants mostly speak a dialect of Mixtec and many of them do not speak Spanish.

Embroidery is one of the longest lasting characteristics of pre-Columbian craft. Artisans learn how to weave during childhood and the practice is essential to the survival of their cultural traditions as textile is considered a substitute for written text. It represents a code written with threads, expressed through symbols, colors, and shapes, and transmits a cultural history with regards to gender, civil situation, and social status. Mostly, weaving with waist looms is done by women, while standing loom is done by men as well.

After these initial visits, I returned to Oaxaca and continued building the team of facilitators and collaborators which I had begun while making connections in the region. In October 2013, I gathered design practitioners that I had known or met while conducting this research for a meeting. This meeting included Maddalena Forcella, Omar Fernández, Fátima Díaz, and Eric Chávez. There, I proposed that we explore the importance of oral tradition in Oaxacan design and production processes. I shared my first impression of my visits to the region and we discussed ways in which we could approach the artisan communities while recognizing oral traditions.

There and at subsequent facilitators' meetings, we would generate a dialogue about topics of co-creation and participatory dynamics, and hear each other's experiences in order to develop our approach and to discuss needs, aspirations, preconceptions, and goals. If some could not be physically present, we held video meetings to share our goals and visions as workshops transformed. We mutually provoked each other into thoughtful discussions which were documented in audio, video, notes, transcriptions, and photos. These often became instrumental to our workshop development and set directions and dynamics on processes.

We facilitators approached the workshops with an interest to understand the existing social design practices in Mexican indigenous communities and also the historically established political parameters in those communities. This called for our reflection on and consideration of previous interventions carried out in indigenous life that resulted in destruction of cultural practices and an altered mode for recording and transmitting cultural traditions.

Cultural artifacts function as tools of connection between people across time and space. For that reason, their value cannot be considered without regard to function or separated from the socio-cultural systems that they belong to. Klaus Krippendorff writes, "The objects not only have orientation relating to the user, but they are also oriented in relation to places in the space built by other objects" (Krippendorff, 1995, p. 165). As such, it was necessary for us to maintain the social and cultural contexts where the practices took place, navigating within their territory while also making obvious the tacit knowledge held by both the facilitators and the artisan communities. Within this context, our task was to transform tacit knowledge into explicit and transferable knowledge while leaving room for heritage as the enterprise of knowledge in the process—a value which Eva Brandt regards as both social and democratic (Brandt et al., 2013; Polanyi, 1966; Collis, 2010; Crouch and Pearce, 2012).

Design Anthropologist Gislev Kjærsgaard (2013) coined the term "knowledge pieces" to refer to artifacts which enable the transfer of individual intangible knowledge to collective tangible knowledge. In this vein, we facilitators considered oral tradition as a "knowledge piece" since, for centuries, it was the only way of relaying and preserving the cultural values and belief in the communities we visited. Through oral tradition alone, these communities generated orientation patterns for conflict resolution and complex symbology systems, which provided both order and meaning.

As tangible artifacts, drawing samples represented the collective imagery of community members and served as a medium for articulation between facilitator and artisan during the workshops. In the oral workshops, women, children, and the elderly from different communities expressed their oral traditions and mythology through drawings which I summarized in a dossier. This way, the presented and represented material was transformed into items of tangible knowledge which could be seen and handled in subsequent workshops.

During the oral workshops, we facilitators recorded images and legends that artisans discussed. Through their patterns, they expressed their own cosmos and vision of the world. The cotton fabric worn by women in San Pablo Tijaltepec displayed motifs containing figures born from the symmetry of nature and abstractions of symbolic elements referring to pre-Hispanic times. Their beliefs and visions about the universe are shared through their oral narrative as they tell their stories that also contain their cultural preferences. Through observing the images and hearing the legends contained in their oral narrative, we found it possible to exchange knowledge.

Prior to the initial workshops, and in order to better understand the role of oral narrative in cultural transmission, I traveled to San Cristóbal de las Casas, Chiapas, and interviewed a Tzotzil woman named Maruch Méndes. As a natural healer, she was a well-respected member of her community; her words and their associated healing powers carried weight. My perception of how Méndes maintained a powerful social and cultural position in her community due to the impact of her words on others reinforced my interest to consider the role of oral history and cultural narratives in design practices.

During the month of February 2014, we facilitators held three narrative workshops in different communities that we had visited in 2013 during the ethnographic phase. All of these artisan communities sought support in their efforts to access new markets and gain greater agency to preserve their traditional crafts. The first oral workshops were held on February 8th at Santa María de Atzompa, and facilitators Omar Fernández and Fátima Díaz joined. As we had agreed in our facilitators' meetings, we approached the community by promoting discussions and collaborative production. This included the community's telling of local stories, personal stories, and legends, combined with their own imagery which was recorded on natural paper I sourced from the Etla Paper Factory.

Oaxaca is well known for its artisans' use of a wide array of materials. During the exploratory phase of the project, I visited a village settled in the mountains named San Agustín Etla. The surrounding mountains there are populated by oak and pine trees, as well as many water springs. An organization there named Coperative Taller Arte Papel Vista Hermosa—or better known since its establishment as La Fábrica de Papel de Etla—is a cooperative using sustainably sourced natural fibers and traditional methods. I sourced paper sheets from them to record the expressive drawings at workshops.

On February 15th, Omar, Maddalena, Fátima, and I held a second workshop in Santa María Yavesía. These artisans were familiar with wood working, and we facilitated a narrative workshop where the community produced drawings on Etla paper with black ink. We were careful to observe the attendees' use of space and their spontaneous grouping of people by age and social roles. These patterns occurred in both communally owned and private spaces.

A week later, on February 22, the team and I conducted a third workshop at San Pablo Tijaltepec. Reaching the village required a long and uncomfortable journey due to poor road conditions. An awkward situation developed when men demand paid compensation for participants as they believed the workshop had been a government-sponsored activity. In this experience, we facilitators learned many new aspects of local politics, including expectations and dependencies on larger governmental campaign strategies.

After concluding this initial set of workshops, the team and I met a number of times to discuss possible dynamics for our workshop practices. Most importantly, we considered how our roles as facilitators could be in improved for our development of future workshops. We wanted to ensure that these workshops were collaborative experiences and not received as a colonizing effort. This required us to be critical of our own positions as facilitators in order to avoid situations that could inadvertently lead to imposition. As design practitioners who appreciate the possibility that design practices may be powerful tools for social change, we did not want to make western assumptions about the communities with whom we collaborated. We made important decisions about the facilitator role, and we agreed to ask general questions to spark more open-ended discussions of culture and society through oral history and design. During our meetings, we also discussed the ways in which we could better link the communities' tacit knowledge to their production of cultural artifacts as a social mediator. To better promote collaboration, we sought to create situations where tools could be shared to experiment and explore ways of producing without a knowable artifact form.

After making these observations, the team and I became much more critical of what role design practitioners should play while attempting to collaborate with artisan communities. We opted to be much more observant of our own roles as facilitators during the workshops. With these group observations, I was able to better record and articulate the significance of our collaborative work. In early November 2014, Laura, Maddalena, and I prepared coming workshops based on samples of Maddalena's work using natural dyes. Working with artisans at San Pablo Tijaltepec, we introduced the natural dye techniques and commissioned their production of physical artifacts based on their drawings created in a previous workshop. As facilitators, we then discussed how we could promote the artisans' transformation of their drawings from 2D samples to 3D objects. From here, we facilitators and artisans shared the goal to produce these objects as artifacts to then be scaled and marketed to a wider audience.

In January 2015, CADA Foundation® was constituted as a platform to facilitate relationships between local artisans and suppliers who wanted to join the global economy and strengthen local presence. With sustainable models in mind, the platform was intended to assist artisan communities in dealing with environmental, social, and economic aspects through a more equitable dialogue. It was also aimed to structure a juridical and administrative body which could agglutinate potential collaborators—including design practitioners and donating partners.

Upon generating more funding for the work, I returned to México in 2015, and we facilitators set out to conduct a new series of workshops with the San Pablo Tijaltepec and Santo Domingo Tonaltepec communities. These workshops were held over three consecutive days beginning with narrative discussions similar to those that we facilitated in previous workshops. On the second day, we engaged in a more general discussion with artisans about artifact production and we collaborated with artisans as they designed and created transformative artifacts. By the third day, attendees had molded figures out of clay based on their drawings that they produced during narrative discussions to then be finalized.

After this new workshop series, I organized another facilitator meeting. We noted how dialogue about craft during the workshops usually transformed into more personal conversations with artisans about their approach to their craft as well as their processes of learning and mastering their handwork. These conversations led to our thinking about the social aspects of design and artifact production in local communities. From our encounters in the workspace, we also found random pieces of materials remaining after the workshops. We worked to generate several prototypes with these new materials, attempting to connect disparate design ideas while also remaining rooted in the design practice and leaving space for discussion and mutual learning. The resulting prototypes display a weaving together of analysis and design through an interactive, hermeneutic process of "task-artifact cycle."[3] Our discussions then developed to consider how materials, volume, and scale of production contribute to the social attributes of artifacts based on a collective body of people and history.

Various studies on the social dynamics of artifact exchanges show the importance of such objects' social functions in representing identity. For this reason, a community's cultural artifacts cannot be relegated to a simple function without also considering identity and social context. As Balaram writes:

> The cultural artefacts and systems acquire meanings integrated with the symbology from which they come. This communal way of participating

> and carrying that meaning, creating the value system, is actually what builds the culture of a community.
>
> (Balaram, 2009, p. 131)

Perhaps the best example of a productive discourse around design emerged from the Ahmedabad Declaration of 1979, which enunciated the role of design in "developing" societies. The declaration was the culmination of a meeting hosted by India's National Institute of Design (NID) in Ahmedabad and was organized under the direction of the United Nations Industrial Development Organization (UNIDO) and the International Council of Societies of Industrial Design (ICSID). Under the coordination of Asoke Chatterjee,[4] this landmark meeting was aimed to understand craft design in India's post-revolutionary years. Following the 1960s creation of the NID and Charles and Ray Eames's "India Report" (1958), North American designers were invited to support India's transition toward an industrialization of their crafts. Their singular mission was not to imitate the developed countries but to research ways to approach local social problems with the help of local knowledge. Design practice was thus suggested as a method of tackling the social demands of India's rapidly changing reality, forming the basis for the founding of the NID (Ranjan, 2011; Ghose, 1995; Balaram, 1995).

I relate the concept of design to the ways in which a produced object acts as an artifact to reflect a particular vision of interaction between people and their respective cultures across time. In this sense, artisans are powerful figures who are able to modify and convey social or cultural preferences in their acts of making. Transforming from individual and un-tangible to collective and tangible through internal narratives of production serves to reinforce collective knowledge. In focusing on the artifact, one must also analyze the context in which it is created, the techniques and materials used, the existing link between artisan and artifact, and also the link between the artisan and their community that derives from their cultural heritage (Ornelas & Gregory, 2011). In such a practice of social design, the artisans' making of the communities' cultural artifacts seeks to favor the exchange of knowledge and values between communities and design practitioners. In the context of equality, this dialogue, as a practice of correspondence, may be adopted by their social, economic and even their political structure.

Through using new materials and forms during workshops, we together created situations between the artifact and social interaction that generated knowledge exchange in a system of correspondence. As artisanal knowledge passes from generation to generation through ways of making, participants in the workshops were able to express their community interests and explore new avenues of making through their cultural artifacts. Through discussion and collaboration, the artisans were able to reflect upon their goals more clearly within a group so that they may create artifacts and realize their strong desire to reach a wider audience. Participation in these workshops was both inclusive and democratic in order to foster a critical working environment for meaningful discussion and realization of community goals. Through building a common community objective together in these discussions, artisans were provided with a space to express agency and were not simply subjected to a predetermined production process. In this way, we felt that our next workshops would prioritize design as a form of energy toward positive social and economic change.

We facilitators organized another series of workshops between June 2015 until March 2016 at Santo Domingo Tonaltepec and San Pablo Tijaltepec. Based on our experiences in the previous workshops, we all set out with clear objectives for experimenting with dynamics for production and market exposure. At one point during this new project phase, the artisans Rosario, Rosalía, Natividad, and Sebastiana asked for additional workshops pertaining to practical business-related issues such as value and pricing, market demand, commissions, and transactions forms. We agreed that it was important to focus on the semantic language conveyed by artifacts within the community in order to assign base market value on context.

At this stage, the work was framed by the process of making things together—collaboratively—involving people with different cultural backgrounds and social systems. As we conducted the workshops, we facilitators weaved our design analyses with artisans' tacit knowledge, and together, we generated a process for creating prototypes that reflected an interactive and hermeneutic process of the task–artifact cycle. Throughout the process, we observed and recorded the artisans' reservations and concerns during production and after the projects were completed.

At this point, the project gained momentum with the teams' full attention toward developing new tool processes for artisans' production of prototypes using several materials and technics. We actively reflected on the process and our experiences during the workshops as artisans developed their goals and prototype ideas in order to ensure that our workshops served as collaborative spaces for discussion and the mutual exchange of knowledge. Through our discussions during the workshop, we hoped that the power of the artifact's narrative could win over its function in the global market. This involved our reflections on the conception of craft and technique used by artisans as they worked to reflect their social preferences and visions of interaction within their culture throughout history. These discussions and reflections helped us as designers to negotiate issues of democratic participation and social hierarchies so that a context of greater equality could prevail in the workshops. Ultimately, we hoped that these sorts of democratic relationships could also develop in their social, economic, and political structures as the artisans more clearly realized their goals and collective vision through these sorts of discussions.

During this final phase, we facilitators scheduled several workshops which I describe in detail in Chapter 5. In these workshops, the San Pablo Tijaltepec artisans transformed their artifacts after we invited a seamstress named Rufina to join our trips to the village. Rufina showed the artisans how to fix a broken sewing machine in the community, and also shared information on finish seams. This involved her sharing with the artisans—Rosalía, Rosario, Natividad, and

Sebastiana—how to turn seams inside-out for flat and even finishes with the machine. Rosalía, Rosario, Natividad, and Sebastiana also organized a collective called Ñaa Ñanga (Mixtec for Toy Women) to present themselves in a communal way for their appearance at the Second Mesoamerican Textile Congress (TEXTIM II) and Expo-Sale at Oaxaca de Juárez. In order to fund production costs for the artisans, I commissioned 25 sewed pieces from each artisan for a total of 100 sewed pieces.

Also, around that time, I ran out of funding and I was not able to continue covering the facilitator team, sample production, or material for future workshops. The artisans' development of a collective, however, did have greater influence in their production, marketing, and sale of their textile artifacts. Our work also began to gain more attention from other interested communities within the region. So, in 2017 we developed a pilot training workshop at the Textile Museum in Oaxaca for local design practitioners and cultural liaisons. I also traveled back to San Pablo Tijaltepec in mid-June to respond to Ñaa Ñanga´s petition for another workshop on sales processes. Eventually, a new workshop at San Pablo Tijaltepec was held in July of the same year.

Meeting Challenges with Reflexivity

As the project took shape over the course of many years, I faced a number of challenges that arose from various sources. The isolation of the communities that requested our collaboration and our lack of funding were important issues. Also, our abilities as a diverse team of designers to grasp cultural differences parallel to personal and professional circumstances were aspects that we were forced to address and reflect upon along the way. I define these difficulties more as experiences and perceived logistical issues related to geographic landscapes and language barriers as part of the process. I also reflected on questions of gender, power, and authority as these aspects were pervasive in the everyday lives of local community members as well as my own reflexivity as a researcher and practitioner.

At one point, the DIF demanded that the project be positioned within a favorable political framework that was far removed from the objective of the project. This demand would severely alter our collective goal not to force an agenda or impose an objective upon the local communities, so I decided to dispense with the aid provided by DIF well into the project which resulted in severe funding challenges. The situation also proved to be a learning experience for us facilitators about local power relations and the authority of organizational and governmental entities in the daily lives of the artisans. Another consequence of our decision to reject the DIF's funding resulted in our founding of the CADA Foundation, whose primary goal remains to encourage and facilitate local artisans and suppliers into becoming part of the local and global economies to this day. By focusing on Social Exchange Through Design, the foundation was born as a platform intended for collaboration between artisans, practitioners, and designers.

My position as a blonde foreigner "güera," in México, is equated with being European or "non-native"—and a woman led me to think more critically about ways in which I could avoid being considered patronizing or colonizing. While I did not intend to force cooperation or modes of modernization, it was important to me that my role was both actually and perceived as that of an observer. This involved me reflecting on my inner views, perplexities, and surprises throughout the project. I also did not want to be perceived or act as an activist, as I am not an agitator. This project aims primarily to generate mutual reflection for communities and practitioners and does not aim to claim rights for marginalized voices. We developed these workshops to provide spaces for discussion and reflection. As facilitators, we sought to consult with and listen to artisans while also fostering critical discussions surrounding former actions to document a more democratic process of producing and preserving communal cultural heritages.

Early on, I recognized sensitivity and empathy as two important concepts for facing issues during the project. It was essential to have a team of local collaborators who could help guide me while we maintained an organic and respectful research design practice. The facilitator team was comprised of designers who were experienced in working with artisans from indigenous communities and who knew—and valued—their ancestral techniques. The work was framed by the process of making things together, among people with different cultural backgrounds, social systems, and core beliefs.

My own personal and professional background served as important factors during this project. I am originally from Galicia—a region in the northwest of Spain—where Galician, a language specific to that region, is spoken in addition to Spanish. We all shared Spanish as a common language—both artisan participants and facilitators—so we were able to communicate easily and directly. Also, in Galician culture, communal and family values remain powerful and relate to food rituals similar to those that I witnessed in Oaxaca. For example, sitting and eating together symbolizes acceptance. My family also has ancestral links with México and Latin America due to migration rooted in changing labor relations. Some of my ancestors and relatives migrated from the region, and I grew up learning about and appreciating their culture and gastronomy. In this case, my rural Galician background granted me a more comfortable understanding of local Oaxacan governing structures, logics, and social roles. As a designer and strategist, I possessed close referents within my cultural background and experiences as a professional that allowed me to navigate more easily in certain cultures.

In the mid-1990s I moved to New York to study architecture and I worked for some years as an architect and designer for major international brands. My tasks mainly focused on creating a global and strategic vision for these companies. These experiences provided me an international pool of resources and contacts contributing to my formulation of a vision and an approach for this project. These contacts and experiences provided me with particular knowledge in terms of marketing, merchandising, sales, and media for the workshops. I also have experience from collaborating with academia and research fields. I have developed my design skills through my professional and personal experiences. The specific aim of my study came into focus through my experiences teaching and elaborating on design studios dynamics. In particular, my experiences leading design research groups aimed at exchanging knowledge between students of diverse cultural backgrounds, ages, and ambitions contributed to my formulation of this present work.

It is important to note that my role as a design practitioner as well as my personal experiences and positionality contribute to my findings. I combined my knowledge as a practitioner and student with action to considering how places, people, and materials come together through the practice of design to develop mutual correspondence. Thus, I believe that my development of ideas based on observation and interpretation of imagined or real situations is critical to this research methodology.

Book Plan Structure

I organize this book in two parts. Part One contextualizes the theoretical and methodological foundation of the project. The first part situates this work in the specific socio-political history of Mexican indigenous communities, and explores the systemic influences of globalization and colonization in research and researcher/indigenous relationships. Part Two, beginning with Chapter 4, tells the story of the project through case studies that were conducted between 2013 and 2018 involving several indigenous communities who practice a variety of craft and created several distinct types of artifacts.

In this chapter, I introduce the objectives of the project and discuss its overall development through time. This involves telling story of how I gathered a group of design practitioners to serve as research facilitators who, along with myself, conducted the project's fieldwork and workshops through a social design practice. Contained in this story are examples of challenges that the team and I faced which were related to key power dynamics rooted in different contexts and our methods for navigating cultural differences throughout the practice. In telling this story, I also introduced some important concepts to the project which I discuss in more detail in Chapter 2.

In Chapter 2, I detail key theoretical concepts of social design to the reader which are activated through examples that I provide in a detailed case study. This chapter situates the research in the context of globalization and colonization, and examines the consequences of said structures on community needs. Chapter 2 also introduces my novel approach which utilizes what I call "proximities of design." This methodology—examined through space-time, making and materiality, and social proximities—defines the ways that designers and non-designer communities can build a shared knowledge and experience in order to facilitate communication around "intangible artifacts." This involves recognizing oral narrative and differing ontologies/cosmologies during the making of prototypes and tangible material artifacts. This methodology builds on existing theories of design-based research, Clifford Geertz's notions of approximation, the works of Elizabeth Tunstall (2013), Luis Villoro (1998), and Homi K. Bhabha (1994), as well as Barbara Adams' theories of correspondence.

Chapter 3 examines in detail the specific cultural, epistemological, and socio-political history of México as it relates to colonization and indigenous peoples. It also explores the history of socio-political factors and relevant scholarship that continue to shape indigenous identity narratives and policies in México, and contextualizes this research project's humanistic approach to the people and objects involved in this research.

Chapter 4 provides a narrative account of the research project starting with ethnographic observation, the development and deployment of several workshops that engaged multiple communities around Oaxaca, and eventual artifact prototyping. This chapter also outlines several pivots, obstacles, logistical challenges, and considerations that are typical for design practice-based research, although often removed from the final reporting of case studies. This chapter details how facilitators navigated their own positionality and power dynamics with indigenous artisans, including successes and failures to navigate proximities of design and correspondence and also demonstrates how the novel methodology proposed was activated in practice. It also outlines outcomes, such as the creation of the CADA Foundation, and the meta-research/documentary that will accompany this book.

Chapter 5 takes the form of practitioners' field notes over the course of the entire project. It shows the transformation of the project and process, including a reflection on methods of communication for social design; the development of collaborative interactions between researchers and indigenous artisans framed by the process of research-through-design and making things together; collaboration between people of different cultural backgrounds, economic systems and core beliefs. This chapter also shows how facilitators exchanged dynamics and materials with artisan participants during "*el-hacer*"—the making.

I conclude the book with a reflection on the research project case studies, its successes, its challenges, and its failures and I suggest my wider insights into the future this design practice and research approach. The conclusion also summarizes the potentials of a novel "proximities of design" methodology, and possible future directions researchers can take in the realms of social design, decolonized design research, and research with indigenous artisan communities.

Notes

1 The DIF is the National System for the Integral Development of the Family, a Mexican public social assistance institution that focuses on developing the well-being of Mexican families, and is spread throughout the country with different locations in each Mexican city.

2 Note: "The making" is a rough translation of the Spanish word "*hacer.*"

3 Concept adapted from *Designing Interaction: Psychology at the Human-Computer Interface* (Carrol et al., 1991). See more on "task-artifact cycle" in Chapter 4 at the end of section 4.4.

4 Asoke Chatterjee was the director of Indian National Institute of Design from 1975 to 1995. He promoted design education in India and is actively concerned with the future of artisans and hand activity both in India and globally.

References

Balaram, S. 1995. Product symbolism of Gandhi and its connection with Indian mythology. In: Margolin, V. and Buchanan, R. eds. *The idea of design*. Cambridge (Massachusetts), London: The MIT Press, pp. 127–143.

Balaram, S. 2009. Design in India: the importance of the Ahmedabad declaration. *Design Issues*. 25(4), pp. 54–79.

Bhabha, H. K. 1994. *The location of culture*. London and New York: Routledge Classics.

Brandt, E., Binder, T. and Sanders, E. B. N. 2013. Tools and techniques: Ways to engage telling, making and enacting. In: Simonsen, J. and Robertson, T. eds. *Routledge international handbook of participatory design*. New York and London: Routledge, pp. 145–181.

Candy, L. 2006. Creativity and cognition studios. [Online]. University of Technology, Sydney CCS Report, 2006-V1.0 November. Available from: www.creativityandcognition.com

Collis, H. 2010. *Tacit and explict knowledge*. Chicago, Illinois: University of Chicago Press.

Crouch, C. and Pearce, J. 2012. *Doing research in design*. London, UK: Bloomsbury Publishing, Plc.

De Avila, A. 2013. Hilo del País de las Nubes. Oaxaca: Museo Textil de Oaxaca.

Eames, C. and Eames, R. 1958. The India report. Paldi Ahmedabad (India): National Institute of Design.

Geertz, C. 1983. *Local knowledge: further essays in interpretive anthropology*. Vol. 5110. Basic Books.

Ghose, R. 1995. Design, development, culture, and cultural legacies in Asia. In: Margolin, V. and Buchanan, R. eds. *The idea of design*. Cambridge, London: The MIT Press, pp. 187–203.

Gislev Kjærsgaard, M. 2013. (Trans)forming knowledge and design concepts in the design workshop. In: Gunn, W., Otto, T. and Smith, R. C. *Design anthropology: theory and practice*. London and New York: Bloomsbury, pp. 51–67.

Hunt, J. 2011. Prototyping the social: temporality and speculative futures at the intersection of design and culture. In: Clarke, A. J. ed. *Design anthropology: object culture in the 21st century*. Viena: Springer, pp. 33–45.

Krippendorff, K. 1995. On the essential contexts of artifacts or on the proposition that design is making sense (of things). In: Margolin, V. and Buchanan, R., eds. *The idea of design*. Cambridge (Massachusetts), London: The MIT Press, pp. 156–184.

Malinowski, B. and De La Fuente, J. 1957. La Economía de un Sistema de Mercados en México. In: Acta Anthropologica, Época 2, 1(2). México D.F.: Escuela Nacional de Antropología e Historia, Sociedad de Alumnos. (Drucker-Brown, S. 1982. Malinowski in México: the economics of a Mexican market system. London, Boston, Melbourne and Henley: Routledge & Kegan Paul.

Ornelas, Y. and Gregory, J. 2011. Cultural artifacts as emotional catalysts and social mediators in design research. 4th Conference on Design Research. Netherlands: ITT Institute of Design.

Polanyi, M. 1966. *The tacit dimension*. Chicago and London: The University of Chicago Press.

Ranjan, M. P. 2011. Design for good governance: information and communication as a platform for change. Keynote Lecture at the 4th International Conference on Contemporary Computing. Jaypee Institute of Information Technology, 9 August 2011, Noida.

Schon, Donald. 1983. *Reflective practitioner*. University of Michigan, New York: Basic Books.

Tunstall, E. 2013. Decolonizing design innovation: design anthropology, critical anthropology, and indigenous knowledge. In: Gunn, W., Otto, T. and Smith, R. C. eds. *Design anthropology: theory and practice*. London and New York: Bloomsbury, pp. 232–250.

Villoro, L. 1998. Estado plural, pluralidad de culturas. México D.F.: Editorial Paidós/ Universidad Nacional Autónoma de México.

Yelavich, S. 2014. Introduction. In: Yelavich, S. and Adams, B. eds. *Design as future-making*. New York: Bloomsbury Academic, pp. 12–17.

2 Framing the Practice

Globalization, Social Design, and Proximities of Design

The concept of "modernity" often relates to the actions taken toward solving social issues to imagine new futures. This idea has been criticized by many for its lack of realism while critiquing society and culture. Ashoke Chatterjee, head of the Indian Institute of Design, criticized the conceptualization of "modern" design aspirations as arriving from "richer lands," arguing that this incursion had no justification because the process itself evolved purely through consumer trends with no consideration of local cultural parameters (Ghose, 1995; Whiteley, 1997). The modernization myth tends to view indigenous societies as "pre-capitalist;" or, as part of the history and economic transformation of Western society. Dating as far back as Adam Smith, Western theorists used ecological—and eventually Darwinist—metaphors to describe the natural history and development of the Western market system. In doing so, those theorists held up indigenous societies as living proof of a singular transformative process that must progress toward a universal modernity. These narratives continue to inform the ways in which indigenous communities are perceived by many as "backwards" in comparison to Western economic history and the current reality. However, many indigenous societies are a-capitalist, not pre-capitalist. These societies conform to an alternative socio-political system which makes little sense when viewed through the lens of Western values, history, and identity politics.

In the globalist discourse which has descended from the ideas of modernism, a new world is expected to center itself around individual needs with no abstract planning for the different needs of diverse social groups. The issue with this global perspective is that it assumes the existence of "tabula rasa" within every nation state and associates cultural, socioeconomic, and aesthetic scales with a predictable and uniform pattern (Ghose, 1995; Acha et al., 2004; Villoro, 1998). Such patterns seem only to arise from a Western capitalist lens; and thus, the nuanced discussion of how preserving cultural differences in the world can lead to positive outcomes is often constrained.

In light of this, I find that pre-made solutions simply do not exist that are expected to work effectively across any or all contexts. Instead, we should approach every sort of specific context at the local and personal level by relying on people's support and by considering the individual and intersecting

DOI: 10.4324/9781003360803-3

structures of specific societies. Likewise, the local challenges, macro-aspirations, cultural preferences, and stories of each community must be considered. Within a non-Western discourse, Social Design practices have been viewed as a possible bolster for more resilient and transformed identities within the context of Globalization as it presents a conflict between ideas of "modernism" and of heritage (Manzini, 2015; McCormack, 2016).

Globalism and Local Community Contexts

One aim of this project is to bring wider attention to, and to raise communities' own awareness toward, the value of their ancestral craft techniques. In many communities, these methods of producing and preserving cultural heritage were and remain at risk of disappearance. Artisans working within indigenous communities in Mexico, India or China have previously maintained a strong resistance to "modernization" efforts by isolating themselves to safeguard their identities and heritage that their crafts hold (Abendroth and Bell, 2016). Today, however, it seems that not even geographical isolation will prevent reach of policies that bring fast change and destructive consequences to local communities and their traditions. New factors such as increased migration from rural to urban environments, a lack of interest from younger generations, and the threat of indigenous policies aimed toward cultural homogenization contribute to this risk. All of these factors are often associated with the idea of "globalization."

The term "globalization" has appeared many times in academic discourses of the past decade, and is frequently viewed as a main driver of change in contemporary societies. Broadly speaking, the term describes a series of policies and processes that transform social and cultural realities at an unprecedented rate. Globalist policies expand the flows of human labor, raw materials, commodities, and information across all areas of the world. From many theorists' perspectives, globalization is nothing but an extension of the "modern era"—or, of contemporary capitalism in a new stage of development—constantly redefining the direction of a nation and its people by de-emphasizing physical and geopolitical boundaries (Giddens, 1990; Novelo, 1976; Villoro, 1998; Franklin et al., 2000). According to Bhabha (1994), the speed of change brought about by forms of "real-time" global communication has increased the pace of idea and image exchanges. This has resulted in a reconfiguration of space and has allowed metropolises such as Tokyo, London, or New Delhi to restructure themselves by making common changes in design, aesthetics, and culture (Kligman, 2007; Bhabha, 1994; Franklin et al., 2000). These common concepts that relate cultural discourse to a global hierarchy seem to be redefining spaces everywhere either by agreement or organically (Bhabha, 1994). These spatial transformations, coupled with the idea that we are experiencing history in "transition" gives rise to questions regarding the relocation of indigenous languages and cultural knowledge.

For some others, "globalization" represents a new concept that serves to substitute the classic idea of a "nation-state" (Appadurai, 1996; Villoro, 1998).

In these views, globalization provides a new framework of values by which we structure our lives as opposed to those provided by a single "nation-state." While critics of globalization tend to focus on consequences that usually revolve around the homogenization and uniformization of diverse cultures around the world, some positive aspects may exist which provide greater opportunities to geographically remote communities. Some recognize globalization as a new tool for people, commodities, ideas, and information to flow and generate a common universe across a vast network of local differences (Franklin et al., 2000; Featherstone, 1995; Appadurai, 1996).

Mexican philosopher Luis Villoro (1998) discusses how globalization has caused a decline in the values and powers of nation-states as the struggle between nationalisms and ethnicities around the world has weakened any particular state's capacity to manage more-or-less homogeneous identities within its borders. On the one hand, globalization may exist as a vast unifying "state" worldwide, and on the other hand, there is a surge in consciousness of identity among peoples that existed previously in such an invented and imposed homogeneity. The alienating forces of globalization have generated a revival of bonds between neighboring communities, and offer an alternative to the "rational" thought of imposed cultural modernity. This experience contributes another meaning to society and our visions of the future: one with potentially more promise and opportunity but which also comes with new and complex demands. While market competition may drive the establishment of bonds within regional economies, it also demands close cooperation in a global network.

It should be noted that this dynamic seems to favor developed—or, more well-established in the existing global network—national markets, thus presenting countries such as Mexico a massive challenge to overcome on the world stage. Globalist initiatives that promote "more developed" Western cultures and aesthetics seem to translate into symbolic economic development. Local communities' adoption of these criteria is also often contradictory to their historical, cultural, and social contexts and can also result in a devaluation of their culture and traditions in the global economy. This is because the global community tends to view the traditional practices and beliefs of these communities as something "of the past," or which has not evolved efficiently in the context of "development". Within this context of globalization, native identities such as those belonging to Mexico's indigenous population risk further marginalization as the pressures to conform to a global standard of understanding mount.

In the Latin-American context, and specifically regarding the roles of indigenous communities there, Villoro (1998) asserts that cultures have always been forced to measure on a scale against a "superior" culture. But, in this moment of transition, Villoro asserts that the waning struggle for resiliency between states can give way to a new cultural vision: the unity of peoples, regions, and ethnicities around the globe in a plurality. On the contemporary conditions of Mexican communities, Villoro claims:

> El reconocimiento del derecho a la diferencia de pueblos y minorías no es más que un elemento de un movimiento más general que favorece la creación de espacios sociales en que todos los grupos y comunidades puedan elegir sus formas de vida, [. . .] La vía hacia un Estado plural es una forma de la lucha por una democracia participativa en todos los ámbitos sociales.
>
> The recognition of the peoples and minorities' right to difference is nothing more than one element of a more general movement that fosters the creation of social spaces in which each and every group and community is able to choose its ways of life ...The path to a pluralistic State is one form of the struggle for a participatory democracy in all the social spheres.
> (Villoro, 1998, p. 7)

Thus, rather than continue to subordinate varieties of cultures to the aims of a singular "superior culture," Villoro believes that we can begin changing relationships between cultures toward a more equitable plurality of interests. Villoro's perspective provides a decolonial vision for our collective futures as long as we are able to recognize this changing moment as an opportunity to collaborate. In light of this, the post-colonial theorist Homi K. Bhabha (1994) claims that it is necessary to generate a new way of thinking in order to reflect on the process of articulating cultural differences. Bhabha defends the idea of "in-between" spaces where the concepts and values of a nation-people can be seen as not so cohesive. Such spaces provide a platform for articulating individuality—whether it be communal or singular—and allow for new forms of identity and collaboration. For Bhabha, constructing the "in-between" is part of the necessary critique of contemporary culture which aims to re-evaluate the concept of community.

Bhabha and Walter Mignolo, two key scholars in the postcolonial design discourse, suggest promoting "cultures of transience." This refers to social cohesion that contributes to collective resilience and serves as "one necessary corrective" to the homogenization patterns generated by globalization. Argentinian sociologist Néstor García Clancini (1977) suggests a similar concept, naming such cultures "hybrid cultures" as they rise in moments of historical transformation. To Canclini, there are no pure or authentic identities; but rather, varieties of cultures that exist within "borderlines" (Canclini, 1977). Consequently, the right to justify and give privileges to an authorized power in "peripheral" countries—as design theorist Gui Bonsiepe (2005) calls them—does not depend on the persistence of the traditions as such. In this work, the preservation of diverse cultures and their persistence through resiliency during a period of globalization is essential. The main focus of the workshops was to collaborate with communities that expressed their desire for support in their efforts to gain market exposure and who were highly marginalized in relation to access for resources within the current global arrangement. Our effort as design practitioners, acting as facilitators, aimed to create avenues for community resilience against powerful forces of change so that the communities and artisans themselves gained greater agency for representation and to realize their own goals against those forces.

While Bhabha's and Mignolo's conceptualizations of communities differ, both scholars agree that we should all avoid conceptualizing the respective "traditions" of communities as being opposed to change. They posit that the notion of "authenticity"—which is usually regarded as a characteristic of indigenous traditions in scholarly work—is all too often used as a justification to fill museums with artifacts (Bhabha, 1994; Mignolo, 2000). This is to say that indigenous communities should not be taken for granted as untouchable and unchanging entities. Members of these communities deserve the right to know how things are done elsewhere and to have that information shared with them. Imposing a sort of isolation on these communities in order to protect their culture and traditions or in order so as to not influence their culture inadvertently is a consequential decision that cannot be made by an outsider to the community. Making such a decision as an outsider can be viewed as a form of marginalization as it refuses the community the opportunity to adopt what they see as beneficial to their goals and their vision.

As workshop facilitators, we discussed how workshops should deal with or "use" globalization to the artisans' advantage. We prepared the workshops while drawing on extensive knowledge about each of the communities' cultures and techniques, but we also shared information with the artisans that was not local or based within their references. Purposely, we exposed the artisans to new techniques, aesthetics, and artifacts that were unrelated to their culture. The facilitators and artisans met while making artifacts to establish a dialogue and a mutual exchange of practices, production processes, and challenges that arose during workshops. this context, the artisanal knowledge of production served a communal purpose for all those participating in the collaborative workshops. The workshops structured transformative environments during the design process to achieve mutual understandings and collaboration between participants. These "in-between" spaces provided the platform for articulating individuality and helped forge new expressions of identity.

In the rural communities, we also recognized how the democratization of communication served as an advantage stemming from globalization. Even in the most geographically isolated and remote places, local communities are able to keep in touch with each other for support and access information through radio and occasionally through the Internet. Although primarily small-scale intercommunication between indigenous communities exists, it allows these communities to remain deeply rooted in place. This does not mean that the contemporary possibilities for wider connectivity are lost on them. For example, within this context, a variety of new and different local activities became possible and in fact arose, such as the Ojo de Agua Comunicaciones in Mexico.[1] Guillermo Monteforte, one of its founding members, worked to introduce the voices of indigenous communities into the national political discourse. His organization brought an experienced team to different indigenous communities so that they could acquire the necessary video technical skills to preserve their intergenerational memories, traditions, and experiences building bridges. They also provided the communities with media sources for the management, training, production, and sustainability aspects of their projects.

Communities can also connect with the broader world to share their knowledge, and to receive electronic payments for transactions or from emigrated relatives who continue to support their families and communities from afar. Thanks to this greater access to communication—through Facebook, for instance—many local communities were even able to learn about and respond to the Fashion Brand Isabel Marant's alleged case of plagiarism.[2] Such forms of communication between communities help to strengthen local identities through recognizing similarities and differences in lifestyles both regionally and globally. It seems that local communities already tend to use technology and some aspects of globalization for their own benefit. Some more positive aspects of globalization, such as fast transportation and instant communication, seem to open a new dimension of freedom for communities which exist in remote geographical areas (Schumacher, 197, p. 53). Thus, geographically isolated localities are not also socially isolated today. The characteristics of interactions between diverse cultures, economies, and societies remain equally important as geographical characteristics. In today's design field, it should serve practitioners—and the communities that they work with—well to respect the legitimacy of social and cultural differences in order to collaborate in an exchange of knowledge and practices.

Social Design Practice

In the last few decades, collaboration between anthropologists and social designers has become more common to lay the foundations for new sorts of projects recognizing the importance of context in design processes (Marcus and Murphy, 2013). Today, many Social Design theory-of-practice discussions revolve around the integration of culture, society, and economics, and the need for culture to be taken as an irrevocable component of design practice (Ghose, 1995; Novelo, 1976; Ranjan, 2011; Manzini, 2015). The implications of design actions and their consequences should be understood as having socio-political transcendence, seeing as how design can both open and close opportunities within the economy through its involvement with artifacts and services (Dilnot, 2005; Yelavich, 2014; Balaram, 1995).

Victor Papanek (1971), Gui Bonsiepe (2005), and Maldonado (1949) are among some of the first to establish a specific discourse on conceptualizing design within contemporary society. These critiques and visions remain the underlying premise of design practice and the designer's role as a provider of services within a system. These critiques also share a concern for creating a responsible design practice. Based on these premises, the concept of social design practice in the context of this specific research should be understood as a service reflected by the action of "making together"; or, a service developed from the community's point of view and that transcends the individual. This is to say that the practice is concerned with developing collaborative interactions framed within a process of making things together through mutual exchange, dialogue, reflective communication, and a democratic decision-making process.

There is a permeability granted to intrusive global factors—i.e., tourists, designers—who engage with culturally differentiated communities. Resilience may serve as a community's reaction to these factors. It is therefore important to consider the ways in which we approach a community so that our intervention is respectful of their socio-cultural visions of the world. Communities have the right to decide what is best for them, what to adopt and how to integrate it, based on their visions of the world and their recognition of communal or individual goals. We may think of social design as a possible tool for building more resilient communities through a respectful correspondence and by making together.

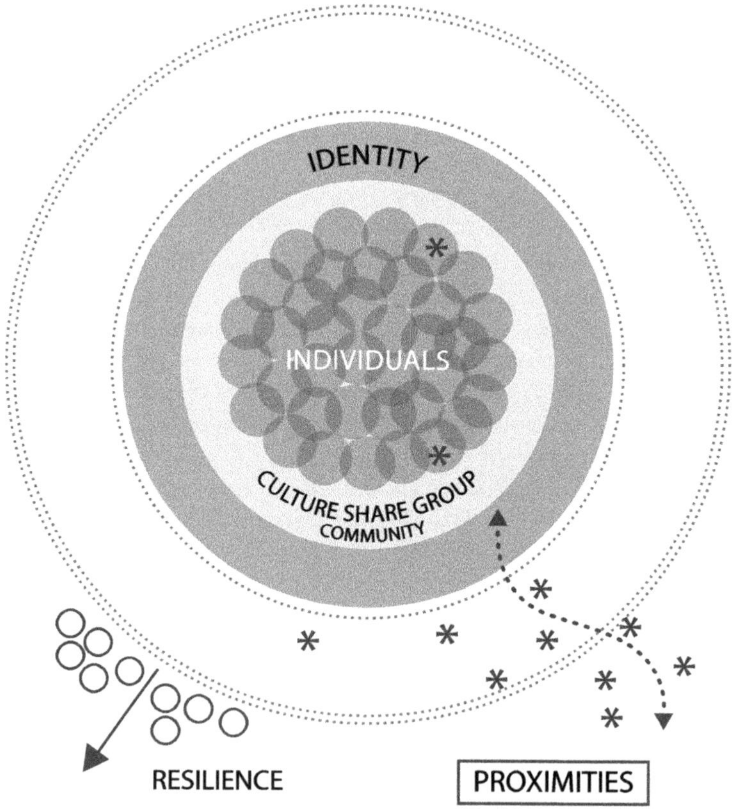

Figure 2.1 Infographic: Communities' Permeability and Resilience. Designed by Carmen Malvar and Rut Martín. CADA Foundation®, 2015.

In our team approximations, we attempted to go further than the purely anthropological or ethnographic studies of persons, places, or materials that could "inform" a process of design. There, we interpreted design as a practice of creating both material and immaterial things. My observations in the field of practice and the values I found there through my decision to ingratiate myself with the local community allowed me to, in retrospect, create a unique space where I could reflect on the fieldwork. In turn, acting as reflective participants, we facilitators were able to respect local agendas, approaches, processes, and symbolisms while also offering artisans an invitation to define their goals and establish new ways to proceed and create together.

Such assessments must be the starting point for building a common objective between design practitioners and a local community. Designers as facilitators, working together with the artisans, shared common objectives and transferred knowledge between each other. The exchange was concerned primarily with the actions, activities, and gestures of correspondences. Our roles as reflective practitioners allowed us facilitators to understand situations and transform workshops through proximities of design established during the practice. To take the deliberate actions and correspondence involved in the project's social dynamic, it was of utmost importance that we, as, facilitators, understood how our different backgrounds—experiences and different ways of producing—influenced our actions and decisions. This practice includes relying on a form of communication for speaking about social design in a reflective way and which is based on an understanding of design after many years of practice experience. In a relationship between knowledge and action, our team's experiences practicing design enabled us to develop interaction and collaboration between people with different cultural backgrounds, economies, and beliefs while simultaneously sharing our knowledges through acts of making of things together.

Our proposals as facilitators were always made while approaching the indigenous communities with humbleness and by reducing the authoritative position that designers have previously taken from within the Western context. Adopting such a role as a researcher and practitioner is critical when people of different cultural backgrounds are involved in a mutual design process. It opens the possibility to find common grounds for all participants during a process that otherwise would not have been possible. Finding equitable participation and distinguishing a unilateral criterion to establish the conditions for the practice are two important aspects of this approach. A facilitator's work should be based on listening first to inform their proposal to transform what is well received into a new design process while also supporting the goals and decisions of the community. It is critical that the humanistic angle which brings all the participants together to exchange knowledge and share values is preserved. The making of communities' cultural artifacts aims to favor an exchange of knowledge and values between communities and design practitioners so that, in a context of equality, this dialogue—a practice of correspondence—may be adopted within their social, economic, and political structures.

Figure 2.2 Making together at workshop. San Pablo Tijaltepec. CADA Foundation®, 2015.

One of the central elements for evaluating and analyzing the impact of an exchange is its reach and duration. Lisa M. Abendroth and Bryan Bell (2016) note that the ability to generate social situations of correspondence over time is one of the essential tools in a social design practice. Furthermore, there must be sufficient time to bind a variety of participants, places, and artifacts and to build layers of meaning and correspondence among them. If facilitators and participants fully recognize this bond over an extended period of time, they may undergo a radical re-transformation and re-consider notions of space and time to subvert the accelerated pace of production set by global market demand. The aim, as such, is to question a taken for granted over-emphasis on the speed of production—a concept which has dominated in these decades of globalization—and to introduce pauses that co-exist with the development of a practice to maintain long-term and more equitable correspondences. In the specific context of Mexico and its indigenous communities, design practitioners must extend temporal expectations and understand the passage of time as a precondition for any deep ties of collaboration (Novelo, 1976; Ghose, 1995).

Based on communities' requests for support in achieving their goals for new market exposure, we facilitators approached indigenous communities found in a state of geographical and economic marginalization in relation to the reach of global capital. In many of the communities we visited, traditional craftsmanship was on the verge of disappearance. Thus, our use of social design practices in the workshops was not merely a means of exchanging production techniques, but was aimed to create spaces for collaborative experimentation and making. The practice of each workshop was aimed at understanding local identity and exploring the ways in which social design could generate socio-economic and political changes that supported the heritage of artisan communities in relation to the global markets. Through social design practice that utilizes a mutual correspondence in these local contexts and with indigenous communities, we—facilitators and artisans—could promote communal identity attributes and agency in order to strengthen each community's position in the globalized world order.

In order to generate this social practice of correspondence, it was necessary to invest enough time to establish a cordial social relationship in which to develop the capacities to engage in open dialogue. The workshops provided a service to bolster resilience for indigenous communities. The workshops were intended to share our tools and knowledge of the practical design field with those indigenous communities that requested support for their efforts to reach new markets. Ultimately, our hope as design practitioners was that the artisans participating in the workshops could incorporate the collaborative work into their own making processes according to their own goals and collective visions. In some cases, we would discuss terms for better resistance, durability, or materiality. In such cases, we worked together and shared different materials such as thread, fabric or clay to explore new artifact forms.

Workshop facilitators also exchanged tools with artisans that are usually related to Western design processes. By combining western tools and design processes with artisanal techniques, artisans could sense their craft efforts in a wider, global context. Previously, outside designers or brands approached artisan communities without considering local culture and without community members' active participation in decision-making about design or production processes. In the workshops for this project, facilitators and artisans exchanged different ideas about design and production through social design practice and with respect to each other's cultural and historical contexts.

When social design is put to practice and subjected to democratic principles, it can become a powerful socio-political activity that gives voice and agency to local people. We facilitators believed that we could not pretend and address this process in only an abstract and global way. We must also address the problems and challenges each community faced in their search for exposure to new markets at the local level, where a better context exists for generating new and more favorable design practices for the global context. This is because deep and committed human relationships exist at the local level which guide behaviors and generate diverse actions and solutions to problems through social exchange and dialogue. A search for these relationships demands a new approach and a new vision adequate to Social Design. We must generate social experiences and establish an equality of values and actions over time so that said relationships can consolidate and blossom. When collaboration happens, each participant should appreciate the time, interest, and dedication of each other. It becomes a matter of trust—the trust of each participant in themselves and in others (Novelo, 1976; Ranjan, 2011; Kester, 2011).

Proximities of Design and "The Making" Together

Design has a long history of commitment to addressing social and cultural issues as well as a history of controversy surrounding intent. In 19th Century William Morris, pioneer of social design in England, rejected forms of mass production and defended the recovery of arts and crafts from a moral standpoint. Morris' movement sought to return value to manual production and to artisans themselves (Pellicer, 1995; Schumacher, 1973). Morris, however, faced a dilemma: on one hand, he was interested in the autonomy of the artisans in creating their pieces; on the other, English high society demanded his pieces without care for workers' social realities (Whiteley, 1997). Such a dilemma remains, perhaps even stronger, in the 21st Century. In the context of our capitalist world of globalized labor, goods are produced in seemingly distant lands where the problems faced by low-wage workers are out of the sight and minds of wealthier consumers. Western design conceptualization has become quite separate from the ideas of knowledge expression and exchange or to the adjustment to human cultures and their needs. This is particularly true in relation to traditional crafts.

Jorge Pellicer presents a case study of Santa Clara del Cobre in his publication *Los artesanos del Porvenir* (*Artisans of the Future*, 1995). The American sculptor James Metcalf planned and conducted the project in a public arts and craft school called CECATI 166 with help from his wife, Mexican artist Ana Pellicer. In their research and artwork, Anna Pellicer and Metcalf applied artisan production processes to support the resilience and cultural heritage of the community[3]. Jorge Pellicer revisits the work that I consider one of the best case studies of documented artifact production in Mexico. Metcalf's writes:

> La producción artesanal es un fenómeno de producción colectiva y puede florecer únicamente en las sociedades integradas igualitarias. Un objeto solo puede estar diseñado propiamente por el artesano, porque el diseño debe ser el resultado del oficio mismo.
>
> (Pellicer, 1995)

> Artisan production is a collective production phenomenon and can flourish only in egalitarian integrated societies. An object can only be properly designed by the artisan because the design must be the result of her own craftsmanship.
>
> (Pellicer, 1995)

My intention in this project was also to maintain the integrity of artisan craft processes and sustain their cultural value. The team and I conduct the workshops to provide artisans with tools that can be incorporated into their craft to the extent that they decide how to adopt new design techniques. Through exposure to these new design ideas and techniques on more equal grounds, artisan communities transformed and bolstered their resilience to pressures of globalism.

During our acts of "making together," we also recognized the importance of promoting the discussion of community values intrinsic to their crafts and goals.

Approaching design as a social practice has helped us facilitators articulate a critical reflection which persists beyond the duration and physical location of a design process during "the making." Craft technique and material exchange among participants are significant aspects of those reflections. All activities were performed in the field, which facilitated the democratic participation of designers and artisans together as a multidisciplinary team. The relationships of trust, the exchange of knowledge, and the techniques used underwent a transformation until paths of common approach were reached to uncover a true exchange between all the participants. It was important that I developed a greater sense of awareness toward this design process experience to develop more adequate ways of framing such experiences in the social realms of action and space—or, through "proximities of design."

"Proximities of design" can be divided into three categories: proximities of design through space-time, proximities of design through making and materiality, and social proximities of design. To understand the context of a community, it's necessary to learn about their value systems and their perception of the world and their place within it. It is important to recognize a peoples' "cosmovision"; or, their vision of the world as it is conditioned by tradition and the activities of social life. This is also related to space and the location of people and objects in it. Space is considered inseparable from subjects or events, as artisans' occupation and navigation through space and relations of power during workshops demonstrated.

The artisan communities that we facilitators worked with during this project also expressed their knowledge through producing objects which also serve as cultural artifacts. The materiality, colors, and patterns of these artifacts represent historical and cultural visions of their community based on their artisanal techniques of making. With this in mind, we facilitated social interactions through the act of "making" by using the artifact as a trigger for opening discussions among all participants. Initially, these conversations focused on

Figure 2.3 Relationships through and within the space. Left Santa Maria Yavesía. Right: Santo Domingo Tonaltepec. CADA Foundation®, 2014–2015.

Figure 2.4 Left: Photo of artisan Juanita Bautista explaining the traditional technique of pepenado. San Pablo Tijaltepec. CADA Foundation®, 2015. Right: Artisan Gregoria Cruz explaining the process of firing clay pieces. Santo Domingo Tonaltepec. CADA Foundation®, March 2015.

physical features and the process of making objects, but later they became about personal experiences and goals.

Individuals reacted differently depending on levels of trust and empathy over time during their involvement in the workshops. Understanding body language involved considering cultural and social realities and positions of artisans in the community. This also involved viewing artifacts and craft techniques as being related to the social, cultural, and political realms of the community. Body language and gestures changed as trust grew, when we changed locations, and as time progressed. Adhering to and respecting the particular social and cultural contexts in each community remained extremely important to recognize various proximities of design.

Through social design, individual and communal characteristics and agencies can be promoted, communities can be strengthened, and solutions to problems can be proposed based on a dialogue of mutual correspondence. Although today the term "social" in the context of design is a broad label applied to those research projects focusing on environmental and socio-political issues, it should be noted that the challenges of social design practice vary between what are often considered "developing" and "developed" countries. Concepts of involvement and time are perceived from fundamentally different perspectives, and issues such as access to resources and local support become more complex as the countries' economic and political contexts come into view.

Social design is a complex and multi-faceted concept which can be approached in many ways. This project was inspired by the social design practice and aimed to demonstrate the possibilities of cultural exchange. But power imbalances are still evident and stand as a roadblock to this goal in every social design project to date (Mignolo, 2000). Designers increasingly seek out collaborations and interventions with places where the culture is vastly different

Figure 2.5 Greetings and sharing food [Digital video]. San Pablo Tijaltepec. CADA Foundation®, 2017.

from their own, pursuing "radical proximity" to the cultures they approach and attempting to achieve new aesthetic categorizations.

In 2010, in his blog on Fast Company, critic and educator Bruce Nussbaum opened a debate on the issues associated with social design practices when he published an article: "Is Humanitarian Design the New Imperialism?" He questioned the interventions of projects known for their use of social design, such as ProjectH and IDEO, which present themselves as projects aimed to resolve human difference through design. Provocatively, he asked: "Are designers helping the 'Little Brown Brothers'? Are designers the new anthropologists or missionaries, come to poke into village life, 'understand' it and make it better – their 'modern' way?" (Nussbaum, 2010, p. 2). Indeed, despite their idealistic motivations to combine the moral calling to do "good" in the world through the power of design, these large organizations routinely fail to respect the value systems of the communities they try to "help" (Nussbaum, 2010).

ProjectH's long and winding project has become a case study in the incoherence of and general detachment from local reality. Nussbaum highlights our need to question the role of idealistic organizations with aspirations for global salvation given the parallels to colonial history. He also forces an important reflection on the impact of initiatives in the field of social design and how to measure it. The cases of ProjectH and IDEO have provided ample reasons for doubt among designers interested in working with contexts where geographic isolation exists, where social interaction is required, and where communities are often limited by scarce economic resources. Such projects spark all sorts of debates and ethical dilemmas, due not only to individual actions but also to misunderstandings linked to cultural barriers, which sometimes produce assumptions that shock and distort the cultural beliefs and social mores of the target communities.

In the case of IDEO and its *Guide to Designing for Social Impact*—a publication resulting from the meeting organized by the Rockefeller Foundation in 2008—Tunstall (2013) criticizes how Western organizations are presented as active agents in the design process, while organizations from "peripheral" countries are presented as passive receptors of interventions. She believes these interventions overlook the design processes—deeply rooted in craft traditions—in those non-Western countries which have allowed local cultural identities to survive the transformational processes of globalization. Tunstall criticizes the imposition of Western thought on non-Western problems and the negligence of ignoring local, non-Western perspectives as information sources. Tunstall claims:

> [The] good intentions by IDEO, bringing design thinking and other non-native principles to India, Africa, or China, for example, risks becoming another form of cultural imperialism that destabilizes and undermines indigenous approaches coming out of other creative traditions.
>
> (Tunstall, 2013: 2)

Therefore, it is necessary to acknowledge that our mere presence—as practitioners—in isolated communities will produce changes even if our participation is calibrated to be respectful. We attempted to foresee these effects through our "approximations" and through "proximities of design" recorded over the course of our workshops. However, since we, as design practitioners, possessed design knowledge that is more related to western practices, it remained vital that we exchanged knowledge in communities with respect for their preferences along with their historical, social, and cultural contexts.

Connected Communities

I have discussed social design as a human process that helps us to understand and interact with the world and each other. In this collaborative work, we facilitators accepted the reality of our globalized world in which communication plays a vital role. We also respect the choice of indigenous communities to keep their distance in order to preserve their culture and traditions. Today, many local communities are forced to perform a balancing act between the information that they possess and that is imposed on them; often, this means balancing between their own interests and the interests of powerful political or economic actors. Rather than subordinate local culture to global design, this research attempts to promote a "culture of transience" governed by local stories and community goals that are triggered through social design and "the making."

In this context, perhaps one of the most important aspects of globalization is how it continuously and radically restructures social spaces and human experiences. Two contradicting forces of globalism work to both open and close opportunities for people, simultaneously. On the one hand, global forces of division create or strengthen existing hierarchies—which translate to unequal spatial and socio-political arrangements—by which communities are besieged, exploited, or isolated. On the other hand, the forces of connection brought by globalization can allow for new exchanges and solidarities between previously disparate places. This connects to the final phase of this project—prototyping and artisanal production—that I outline in Chapter 1.

During the third workshop held at San Pablo Tijaltepec in 2015, we shared with participants a series of black and white images of handmade toys based on the idea of toys as accessible artifacts for all. Some of these belonged to Asian cultures, others to different Latin American regions. In the Santo Domingo Tonaltepec community, we also shared images of utilitarian elements such as decorative objects transformed in other contexts for utilitarian uses. Artisans' observations raised questions and generated a subsequent dialogue. After observing these photos, we asked artisans to choose their favorites. Intuitively, they mostly selected images of artifacts resembling the aesthetics of their own cultural traditions. After their selections of the images, a dialogue between the artisans and the facilitators was established to represent a starting point on how to evolve and structure the ensuing workshop. By sharing these images, we facilitators sought to navigate global popular culture at the local level with artisans through a universal design language, and as discussions were triggered through "*the making*" together.

Toward the final stage of the project, San Pablo Tijaltepec's female artisans also recognized their goal to exhibit their artifact products at the Expo-Ventas[4] organized by the Textile Museum in Oaxaca after attending the workshops. For the occasion, the women organized a collective called Ñaa Ñanga for communal representation in their first exposure to a new market. At first, they were critical of the experience because they did not like the food or the city very much, and expressed concern about the percentage requested by the organizers of Expo-Ventas for their participation. However, artisans were able to sell their blouses at higher prices than usual, and the sales process was transparent. Facilitators developed a specific workshop to explain the details of the event to artisans and to hear more of their concerns.

After their participation in Expo-Ventas at the Textile Museum, Ñaa Ñanga also realized the opportunity to manage the group's communications through social media and manage their own appointments using a mobile phone. Rosalía, the youngest of the craftswomen, asked Laura Quiroz to teach her how to use tools such as Facebook to manage the group's sales on the Internet. It was by the community's own interest and their exposure to these medias' existence that prompted the artisans to view them as positive tools to use in their favor.

In no case was this decision planned or considered in the workshops, but rather through the community's contact with commercial sales activities at the event. These experiences also promoted the Collective´s interest to learn about them, in order to be autonomous and independent in their use.

This would not be possible under a western perspective within which indigenous communities are held "encapsulated" by their role as traditional and locally producing artisans. Under such circumstances, artisans are often excluded from participating in the wider market processes to commercialize their products, from making decisions for their own benefit, or from being able to connect with other realities outside beyond their own community. This "encapsulation" limits the community's capacity for resilience should an intervention arrive to transform their practices and culture in the name of globalism. In short, the lack of a more equitable relationship with the wider world leaves them vulnerable to ill-guided globalist policies.

In this light, it seems that technological globalization may support small communities by connecting them with ideas beyond the limits of their own cultures. A network of social connections allows communities, wherever they are, to generate alliances that transcend the local struggle for rights and values. Moreover, these connections help to bolster community identities and contribute to their continued existences as they face a moment of history "in transition." In sum, the result of globalism's dual forces is a "new space:" one which requires great risks to construct, but which can also offer great socio-economic potential (Arnaut, 2011).

For the artisans, experiences gained through their participation at workshops allowed them a new understanding of culture and globalism and that they can participate there on their terms. Ultimately, they made the decisions to change aspects of their craft or not based on their cosmovision, culture, and communal interests.

Framing Democratic Design Environments

In this chapter, I situated my project within the context of globalization and have examined a few consequences. I highlighted how the feelings of alienation resulting from globalization have, to some extent, promoted the revival of community bonds through more localized communications. Despite this, the possibilities for socio-cultural homogenization produced by globalization still risk marginalizing native identities—such as Mexico's indigenous population—and efforts to achieve resilience in these communities remains necessary.

In 2007, following a series of reflections on design in a social context, the attendees of the second Ahmedabad meeting proclaimed the role of design in developing countries. Among the countries were Russia, Germany, the United States, Japan, Argentina, Brazil, India, and Pakistan[5]. Intellectuals such as Papanek and Bonsiepe also participated in the meeting. They called for design to generate a value system consistent with the society of each "developing" country, but they also criticized the hasty shedding of identity viewed as a necessary change for participating in the global economy (Balaram, 2009; Ghose, 1995). The Ahmedabad Declaration emphasized a number of needs:

> (1) Understanding the values of one's society and then defining a quality of life within its parameters; (2) seeking local answers for local needs by using local materials and skills, while making use of advanced science and technology; and (3) creating new values, addressing priority needs, and preserving plural identities.
>
> (Balaram, 2009, p. 1)

Globalist efforts tend to decontextualize the value of what is "local" and attempt to integrate it into a homogenous global culture that follows a Western model. This process takes the ideas of Western capitalism for granted as a natural vision in all societies. In that perspective, the process also involves the devaluation of "local" traditional practices and designs in order to bolster the value of what is "global"—or more efficient. Rather than consider the complex needs of diverse social groups across a variety of contexts, the global effort to bring "modernity" to remote communities focuses on solving individual problems through competition.

Based on the priorities listed above, the facilitators and myself approached each community with the hopes of having more organic discussions and to develop meaningful relationships. By using our knowledge of design and by learning about the histories of the communities, we were able to frame our experiences through "proximities of design" within a more equitable and democratic environment. Each workshop was thus different and evolved during its development as events unfolded without a forced objective or goal. We employed a social design practice to hear the needs and goals of artisan communities and provided a service to provide support in their efforts to achieve goals rooted in their changing visions of the world.

Maintaining cultural differentiation was a key element in this project. Our aim as facilitators was not to impose a specific methodology or goal to achieve during the workshops. Instead, we valued an exchange of the diverse knowledge that artisans and the team could offer each other. Facilitators and collaborators that took part in the process were designers, mainly Mexican born or had lived there for a long time, they had previously worked with indigenous communities and they had different backgrounds and international training. These characteristics of the team were key to getting into practice the established guidelines to avoid a westernized colonizing approach. Together, on more even grounds, we learned how to share aesthetics and design techniques tailored to the communities' contexts and through "proximities of design." Ultimately, artisans got to decide the ways in which they might change their ways of doing things—if at all. These decisions cannot be made from the perspectives that exist beyond the community and which do not consider specific local context. In this project, we aimed to establish an environment wherein artisans were able to make these decisions and collectively realize their goals. Through our social design practice, we supported the artisans as they built resiliency and bolstered community bonds to stand toward the daunting pressures of globalism.

Notes

1 Ojo de Agua Comunicaciones is a communication project that seeks to promote equity and equality between women and men of indigenous communities by eliminating of forms of violence through the construction of more democratic and egalitarian societies. http://ojodeaguacomunicacion.org/.

2 In 2015, a community in Oaxaca, Santa María Tlahuitoltepec, presented court allegations at national level against French fashion designer Isabel Marant, accusing the firm for plagiarism of their traditional xaamnïxuy blouse. Marant had included a blouse with a very similar pattern in her Spring 2015 Collection Etoile. She was subsequently accused by a competing fashion house, Antik Batik, for imitating a pattern they had included in Bartra for a collection presented in 2014. Marant then faced the Paris High Court in France. While the Oaxacan community did not file for monetary compensation from Marant, they demanded acknowledgment. The Court ruled that neither fashion designer had rights to traditional textiles and noted Marant´s acknowledgment of having "received inspiration from traditional blouses of Santa María Tlahuitolpec, in the Province of Oaxaca, México." The ruling proposed an indirect dialogue with the community through the Mexican embassy in Paris in order to plan an exhibit in Paris about the community´s work and culture. To my knowledge, this exhibit has still not taken place. Eventually, the Tlahuitolpec community and others requested a Certificate for Patrimonial Protection from the INE (Institute of Indigenism) and was written by Carmen Castillo. The case has raised awareness for the cultural—and monetary—value of traditional craft, and has opened legal discussions about the protections for collective property rights.

3 I document the legacy of CECATI 166 at Santa Clara del Cobre, Michoacán, México, and CECATI 133, at San Cristóbal de las Casas, Chiapas, through my visits to these locations and interviews with their surviving Directors: Ana Pellicer, Luis Morales, Marina Sánchez and other of Metcalf's close collaborators.

4 Expo-ventas are trade exhibits organized around different topics, usually involving food and craft sales and craftmanship demonstrations. In this case, the event was organized by the Museo Textil de Oaxaca on their premises or at other local cultural institutions. Normally, organizers will supply space and logistics and are paid a participation entry fee or an agreed upon sales percentage. In the local context, craft sales were usually conducted at open-air markets, prices were subject to negotiation, and neither logistical support nor education activities (such as craftmanship demonstrations) were involved.

5 The conference was sponsored by UNESCO, UNEP, and the government of India. The host was the Centre for Environmental Education in Ahmedabad. This declaration was written in the context.

References

Abendroth, L. M. and Bell, B. 2016. *Public interest design practice guidebook: seed methodology, case studies, and critical issues*. London, New York: Routledge.

Acha, J., Colombrea, A. and Escober, T. 2004. Hacia una Teoría Americana del Arte. Buenos Aires: Ediciones del Sol.

Appadurai, A. 1996. *Modernity at large: cultural dimensions of globalization*. Minneapolis: University of Minnesota Press.

Arnaut, K. 2011. The manifestation for the re-annexation of Belgium to Congo: reconfiguring (post)colonial space. In: De Cauter, L., De Roo, R. and Vanhaesebrouck, K. eds. *Art and activism in the age of globalization*. Rotterdam: NAi Publishers, pp. 130–139.

Balaram, S. 1995. Product symbolism of Gandhi and its connection with Indian mythology. In: Margolin, V. and Buchanan, R. eds. *The idea of design*. Cambridge (Massachusetts), London: The MIT Press, pp. 127–143.

Balaram, S. 2009. Design in India: the importance of the Ahmedabad declaration. *Design Issues*. 25(4), pp. 54–79.

Bhabha, H. K. 1994. *The location of culture*. London and New York: Routledge Classics.

Bonsiepe, G. 2005. Design and democracy. Santiago de Chile: Metropolitan University ofTechnology (English abbreviated translation of an acceptance speech in Spanish of a honorary doctorate). In Bonsiepe, G. 2006. *Design issues*. Massachusetts, USA: MIT Press, 22(2), pp. 27–34..

Dilnot, C. 2005. Ethics? Design? The Archeworks Papers. Chicago: Archeworks, 1(2).

Featherstone, M. 1995. Undoing culture: globalization, postmodernism and identity. Thousand Oaks, California: SAGE Publications.

Franklin, S., Lury, C. and Stacey, J. 2000. *Global nature, global culture*. London, Thousand Oaks, New Delhi: SAGE Publications.

García Canclini, N. 1977. Arte Popular y Sociedad en América Latina. México, D.F.: Editorial Grijalbo.

Ghose, R. 1995. Design, development, culture, and cultural legacies in Asia. In: Margolin, V. and Buchanan, R. eds. *The idea of design*. Cambridge, London: The MIT Press, pp. 187–203.

Giddens, A. 1990. *The consequences of modernity*. Stanford, CA: Stanford University Press.

Kester, G. H. 2011. *The one and the many, contemporary collaborative art in the global context*. London: Duke University Press.

Kligman, A. 2007. *Brandscapes: architecture in the experience economy*. Massachusetts, USA: The MIT Press.

Maldonado, T. 1949. El diseño y la vida social. In: Boletín 2 del Centro de Estudiantes de Arquitectura de Buenos Aires. 2(October-November 1949), Buenos Aires, Argentina, pp. 7–8.

Manzini, E. 2015. *Design, when everybody designs: an introduction to design for social innovation*. Cambridge, London: The MIT Press.

McCormack, C. 2016. Talking heads. Unpublished. Lecture, Southampton University, UK.

Mignolo, W. 2000. *Local histories/global designs: coloniality, subaltern knowledges, and border thinking*. Princeton and Oxford: Princeton University Press, p. 302.

Murphy, K. M. and Marcus, G. E. 2013. Epilogue. In: Gunn, W., Otto, T. and Smith, R. C. eds. *Design anthropology: theory and practice*. London and New York: Bloomsbury, pp. 251–268.

Novelo, V. 1976. Artesanías y capitalismo en México. Tlalpan, D.F.: Instituto Nacional de Antropología e Historia.

Nussbaum, B. 2010. The new imperialism. MIT IDEAS Global Challenge. [Online]. [Accessed 8 July 2010]. Available from: http://mitpsc.mit.edu/globalchallenge/?p=417

Papanek, V. 1971. *Design for the real world: human ecology and social change*. 2nd ed. Chicago: Academy Chicago Publishers.

Pellicer, J. 1995. Artesanos del Porvenir. México, D.F.: Secretaría de Educación Pública, Universidad Autónoma Metropolitana, Ciudad de México.

Ranjan, M. P. 2011. Design for good governance: information and communication as a platform for change. Keynote Lecture at the 4th International Conference on Contemporary Computing, Jaypee Institute of Information Technology, 9 August 2011, Noida.

Schumacher, E. F. 1973. *Small is beautiful: a study of economics as if people mattered*. London: Vintage Books.

Tunstall, E. 2013. Decolonizing design innovation: design anthropology, critical anthropology, and indigenous knowledge. In: Gunn, W., Otto, T. and Smith, R. C. eds. *Design anthropology: theory and practice*. London and New York: Bloomsbury, pp. 232–250.

Villoro, L. 1998. Estado plural, pluralidad de culturas. México D.F.: Editorial Paidós/ Universidad Nacional Autónoma de México.

Whiteley, N. 1997. *Design for society*. London: Reaktion Books.

Yelavich, S. and Adams, B. 2014. eds. *Design as future-making*. New York: Bloomsbury Academic.

3 Articulating Cultural Differences

Indigenism, Politics, and Social Context in México

Various social and political factors have shaped the sorts of policies and narratives that are now changing the status and character of indigenous communities in México. After the Mexican revolution, powerful organizations and individuals have challenged the futures of indigenous cultures and people. New nationalistic policies have aimed to homogenize the diverse character of Mexican society and its many cultures. In this chapter, I discuss socio-political trends in México and their effects on indigenous identities. I will also discuss the possibilities for preserving a diversity of cultures through a humanistic approach to people and cultural artifacts of indigenous communities. Foreign design processes and methods can be adopted to suit the internal vision of indigenous communities and provide a platform for the transmission of group values and beliefs through their cultural artifacts.

Imagined Communities

In México's situation, it is important to note the state has consistently attempted to generate a single cultural identity and aesthetic as an instrument for post-revolutionary nation-building. Through the exaltation of indigenous roots and folk traditions, the government sought to combine pre-Hispanic techniques, patterns, and materials with modernist trends (López, 2010). Critics of Mexican post-revolutionary culture and society, like anthropologist Bonfil (1987), establish a critical discourse regarding the indigenous politics of the moment. They make obvious the clear dichotomy of realities in social structure—represented by a cultural elite increasingly distanced and retired from the reality of the Mexican people and their racial and cultural contradictions.

In this context, Benedict Anderson (1983) defines the concept of "nationalism" as a sovereignty based on the idea of an "imagined community" existing in the "homogeneous empty time" of modernity and progress. But, following the reality of a colonial period, one must avoid fetishizing suppressed cultural identities by fixating on their original form. Such a fixation on indigenous cultures can cause homogenization through a modernist Western prism. Previously, the western expropriation of indigenous cultural characteristics has forced a process of self-objectification through the Western perspective. It is

DOI: 10.4324/9781003360803-4

necessary to generate a new way of thinking that goes beyond origin and subjectivity and which reflects on the process of articulating cultural differences effectively. Metaphorical movements of time and space require a narrative that also moves between cultural and social formations without a local causal center.

Bhabha argues that such a narrative should have its own authority. The codes and languages of this narrative should go beyond a horizontal and sequential cultural revision and giving authority to a "non-sequential energy of lived historical memory and subjectivity." Bhabha (1994) claims that "[we] need another time of writing that will be able to inscribe the ambivalent and charismatic intersections of time and place that constitute the problematic 'modern' experience of the Western nation." Both Fanon and Bhabha, in challenging the pervasive colonial epistemes, defining knowledge as Plato did: "justified knowledge as truth", opposed not only the lineal direction of Western history but also the temporal thread that history has established as an ordering criterion (Fanon, 1967; Bhabha, 1994). Bhabha praises the provocative work of Mexican activist and artist Gómez Peña, who explored the dilemma of identity within peripheral cultures, reflecting on the concept of transculturality and the intersubjectivity it provokes through a hierarchy of differences. Peña made obvious the common experiences of those communities with similar interests and values within the post-colonialist discourse (Bhabha, 1994).

The discourse surrounding identity in México is unavoidably linked to indigenous politics. Such historical analyses have been based on several key studies—some of which remain unpublished in English—in social anthropology. The studies that I summarize below are vital for establishing political and cultural actions in the fields of arts and design, and which form the social basis and scope of this project. It is also important to point out that the post-revolutionary period of Mexican history reviewed here continues to influence attempts at generating a new single culture through reinterpreting the patterns of indigenous cultures; a homogenizing effort that has remained unsuccessful since the time of the Spanish Conquests between 1519 and 1521 (Aguirre Beltrán and Pozas, 1973).

Indigenism and Anthropology in Post-revolution México

It is important to contextualize contemporary México with respect to the nation's history, its diverse cultures, and the cosmovision of its original inhabitants. Tales of conquest and revolution have contributed to the development of México's culturally heterogeneous society which has faced nationalistic pressures toward homogenization. Politicians and other powerful agents of change continue to transform México's national identity through national policies and by constructing narratives that promote less diversity.

México's independence from Spain in 1821 put two opposing forces into play that attempted to confront the indigenous world. The first perspective promoted a romanticized story of indigenous history as a reaction against the previous governing class and their goals to cultivate foreign connections. The second perspective viewed indigenous heritage as a bolster for promoting a nationalist spirit. It was the latter that shaped the indigenous narrative and policies in México, as well as the development of anthropological study. According to Villoro (1998), a group of intellectuals, educators, and politicians imposed their wish for a nation-state—or, an "ethnicized and culturally unified modern nation"—onto the multicultural landscape of post-revolutionary México (López, 2010). The desire and efforts to synthesize various regional cultures and beliefs into a single "representative" form were managed and produced by the dominant powers of society. As such, this effort of more powerful agents involved imposing their particular standards over the rest of México (Villoro, 1998). Considering México's heterogeneous social and cultural reality, this was a complex task. For instance, México has 69 co-official languages, out of which 68 are indigenous and pre-Columbian with 364 variants in addition to Spanish.

In 1864, Francisco Pimentel—a writer, historian, linguist, indigenist, and academic—was already discussing the distinct realities of two different, and to some extent opposed, peoples (Villoro, 1998). In anthropologist Gonzalo Aguirre Beltrán´s words:

> El individuo [. . .] no se siente pertenecer sino a su comunidad, a su linaje y a la tierra del linaje. No tiene conciencia de ser solamente un fragmento de un grupo étnico, más amplio que la simple sociedad nuclear cuando ese grupo excede las dimensiones de una comunidad. Mucho menos se considera miembro de una nación, abstracción que rebasa el mundo conceptual que abarca la estrecha cultura que tiene por patrimonio.
>
> (Aguirre Beltrán and Pozas, 1973, p. 42)

> The individual [. . .] does not feel her belonging but to her community, her kin, her kin´s land. She has no conscience of being only a fragment of an ethnic group, wider than the simple nuclear society, when that group exceeds the dimensions of a community. She even less considers herself as a member of a nation, an abstraction that exceeds the conceptual world that encompasses the narrow culture she has as heritage.
>
> (Aguirre Beltrán and Pozas, 1973, p. 42)

Indigenism, an ideology popular among many Latin American countries during the 20th century, perhaps held its strongest foothold in México. Since its incorporation into the 1917 Constitution, Indigenism guided the relations between the governing class and indigenous peoples. What distinguished indigenists was their desire to celebrate indigenous culture as a part of the nationalistic narrative while simultaneously working on behalf of the nation-state government to homogenize, assimilate, and acculturate indigenous populations. To prominent advocates of the ideology such as Gamio (1916), indigenism had to be "integral"—not only permeating government policies and institutions, but permeating cultural policies and expressions as well.

Nearly every indigenist anthropologist references Manuel Gamio and his 1916 book *Forjando Patria* (*Forging a Nation*) as the first practical implementation of anthropological science in México and a founding document for Mexican indigenism. As Mexican anthropologist Julio De la Fuente declared in 1964, Gamio—influenced by the North American school of Franz Boas (1927)—made a series of remarkable additions to the social sciences and stood out for his compelling case that anthropological knowledge could be a precedent for social action (De la Fuente, 1964). For better and for worse, according to Bonfil (1987), Gamio's ideas "have guided the indigenist policy until very recently".

Written during the tumultuous ten-year revolution (1916–1926), *Forjando Patria* was a manifesto for a Mexican national anthropology as well as a monograph on the nation's complex cultural politics. Portraying the complex—and in retrospect problematic—position of indigenists, Gamio declared that "knowledge of the population is basic for the achievement of good government" (Gamio, 1916). This premise, which was actively applied until the 1970s, formed the basis for an indigenist policy (Bonfil, 1987). Implicit, of course, is the notion that attempts at understanding and dignifying indigenous peoples are valuable only insofar as they are instruments of nation- and ethnicity-building. Gamio wanted to transform the indigenous minorities—whom he saw as foreigners in their own country—into Mexicans (Gamio, 1916).

Gamio established the parameters for anthropological research as a method of confronting the social problems of indigenous groups and also conducted perhaps the first reassessment study of the indigenous people and their cultural symbols (De La Fuente, 1964). He called for the decisions toward and appraisal of indigenous peoples to be conducted from an integral perspective that considered biological, psychological, social, and cultural factors. In his first documented study of the Teotihuacan Valley, Gamio used a variety of analytical methods to consider various aspects of social life in order to apply anthropology beyond the purely academic spheres. He asserted that the cultural manifestations of the population could not be valued from a single perspective and measurement, but they should rather be evaluated and measured within their own contexts. He also urged against the use of standards of superiority and inferiority in comparisons across contexts. Nevertheless, *Forjando Patria* declared that the ultimate goal of indigenist anthropological research was to identify which indigenous cultural traits were worth preserving for the purposes of building a new and unified nation-state.

Often under Gamio's tutelage, several prominent figures in post-revolutionary México continued the work of integral indigenism in anthropology as well as in politics and education. Shortly after the end of the revolution in 1920, politician and educator José Vasconcelos began educational reforms to fight illiteracy by increasing the number of public schools and by promoting greater rural education (Aguirre Beltrán and Pozas, 1973). As the Secretary of Public Education between 1921 and 1924 and as the founder of the Department of Indigenous Culture, Vasconcelos appointed traveling teachers to study the Mexican peoples (Aguirre Beltrán and Pozas, 1973). In his 1925 book *Raza Cósmica* (*Cosmic*

Race)—which set out a program and vision for a unified mestizo Mexican race—Vasconcelos showcased the utopian vision of post-revolutionary nationalists (López, 2010). Moisés Sáenz's book *México Íntegro* serves as another invaluable indigenist reference in the current discussion of multiculturalism and nationalism in México. [1]. Sáenz (1939) shared Gamio's integrationist vision, and the combined influence of their collective work guided the approach of cultural interventions during the 1920s and 1930s; aesthetics and structure of Mexican cultural identity remained at the forefront of discussions about indigenous populations (López, 2010). Gamio's influence in the field of anthropology also extended across México's northern border. In the 1930s and 1940s, under Gamio's influence, American anthropologist Robert Redfield developed the trend of researching the knowledge of indigenous peoples by looking at them from different angles. In the Mexican states of Morelos and Yucatán, Redfield embarked on ethnographic, linguistic, archaeological studies in addition to the traditional physical and cultural anthropology of his forbearers.

Similarly, in 1957, the Polish Mexican team of Bronislaw Malinowski and Julio De la Fuente published their anthropological work on market systems in the Oaxaca Valley. Their study of commercial relationships between neighboring villages and states is still used as a reference for field research seeking to combine social, anthropological, and economic perspectives. Malinowski, oftentimes considered the founder of social anthropology and the first to incorporate theories of psychological functionalism into such work, arrived in México around 1930. He was initially attracted there by the social transformations generated by the revolution. Malinowski's legacy is still often defined by his insistence on anthropologists' immersion into local cultures and language practices. The organizing feature of this method was the navigation of narratives. Malinowski moved from one narrative to the next; following the relationships and interdependencies framed by them.

Malinowski and De la Fuente swore to a consistent and reciprocal relationship between observation and theoretical analysis in their work. They believed it was imperative to collect materials in some systematic fashion and along some standard principles so as to maintain the value of their work. In Oaxaca, they departed from the purely aesthetic descriptions undertaken by other indigenists who, up to that moment, had simply placed cultural artifacts on a pedestal through the lens of popular culture. Instead, they related their descriptions to market principles: consumption, distribution systems, transportation logistics, communal forms of property, and buying power. Malinowski, like many others before him, also showed a concern for social reform. In his work, he discussed the successes of projects such as the artisan cooperatives in Santa María de Atzompa and efforts to raise the working and living standards of Mixtec palm hat weavers[2].

Malinowski's work in México was published in Spanish 15 years after the English publication. Julio De la Fuente helped to translate the original report; however, this initial translation did not include the appendix which contained black-and-white photographs taken by Malinowski and De la Fuente during their fieldwork in Oaxaca. Mexican students of their work lacked access to these photos for decades after the Spanish edition of the work was published. Sarah Drucker-Brown, who published the 1982 edition of the book, was the first to appreciate the value of Malinowski's ethnographic visual material which captured cultural identity and markets, as well as their relationship with artifacts in the context of México.[3]

Alfonso Caso—once the director of the Instituto Nacional de la Antropología e Historia (National Institute of Anthropology and History)—represents yet another indigenist anthropologist who rose to prominence a generation after Gamio. Notably, he unified the various disciplines constituting the "Science of Man," which up to that point were scattered in different private and public entities without clear objectives to guide their work, and without inter-coordination. In the anthropological tradition of Gamio, Caso considered the study and knowledge of the past important only as a means for modifying the present. In his analysis, Caso established the concept of community and group consciousness as the key to defining indigenous persons. To him, the term "Indian"[4] referred to:

> [. . .] todo individuo que se siente pertenecer a una comunidad indígena; que se concibe a sí mismo como indígena, porque esta conciencia de grupo no puede existir sino cuando se acepta totalmente la cultura de grupo; cuando se tienen los mismos ideales étnicos, estéticos, sociales y políticos del grupo; cuando se participa en las simpatías y antipatías colectivas y se es de buen grado colaborador en sus acciones y reacciones. Es decir, que es indio el que se siente pertenecer a una comunidad indígena.
>
> (Caso, 1953, pp. 145–181)

> [. . .] any individual that feels she belongs to an indigenous community; that thinks of herself as indigenous, because this group consciousness cannot exist unless the group's culture is completely accepted; when the group shares the same ethnic, aesthetic, social, and political ideals; when the collective sympathies and antipathies are shared, and she is voluntarily a collaborator in the actions and reactions of the group. That is to say, an Indian is whoever feels she belongs to an Indian community.
>
> (Caso, 1953, pp. 145–181)

In 1940, Caso founded the Instituto Nacional Indigenista (National Indigenist Institute, or INI) with the goal of providing a basis for the anthropological methods to support regional and state projects in indigenous communities. The INI shifted its focus of interventions in indigenous communities away from solely education and local markets and instead toward integral action; or, a more holistic and balanced approach jointly implemented across many societal and cultural structures. Caso was, in some sense, realizing Gamio's idealistic vision for integral indigenism decades later. This required extensive knowledge of the specific lifestyles of each community, as well as continuous study of the changes induced by the methods used. This vision informed subsequent steps in the development of an integral policy, which came into place in stages following the founding of the INI. Below are some notes on his plans for literacy:

> El entrenamiento de estos jóvenes consiste en cursos intensivos sobre alfabetos, metodología, traducción de textos indígenas y propaganda [. . .] El empleo de dibujos con la indumentaria y usos de la región permiten dar a estas cartillas un contenido local. Los veinte jóvenes así preparados forman misiones alfabetizadoras jefaturadas por un técnico en lingüística y trabajan, desde el primer día, con niños y adultos.
>
> Los fines de la educación nacional [. . .] están orientados hacia la formación y capacitación de las generaciones jóvenes para que puedan realizar, en un futuro próximo, la modernización del país; la educación informal, así como sistematizada que se imparte en las instituciones educativas, están empeñadas en preparar los cuadros de técnicos capaces de alcanzar dicha meta. Además, el sistema educativo persigue otro fin: la unificación social y cultural del país para integrar la nacionalidad mexicana. Ninguna otra institución ha actuado con tan clara conciencia en la realización de este empeño como la educación; y el Estado, al fin, se ha decidido a llevar la escuela hasta las ayer inaccesibles comunidades indígenas …
>
> (Instituto de Alfabetización en Lenguas Indígenas, 1945, pp. 140–160)

> The training of these children consists of intensive courses on alphabets, methodology, translation of indigenous texts, and propaganda [. . .] The use of drawings depicting regional apparel and customs affords local content to these booklets. These twenty youngsters are thus prepared to set up literacy missions, headed by a specialist on linguistics, who works from day one with children and adults.
>
> The goals of national education [. . .] are oriented towards the training and qualification of the new generations so that they can carry out, in the near future, the modernisation of the country; informal education, as well as the systematised education in educational institutions, are charged with preparing cadres of technicians capable of achieving this goal. In addition, the teaching system pursues another goal: the social and cultural unification of the country, to assimilate the Mexican nationality. No other institution has acted with such clear conscience in the achievement of this goal, as education; and the State, finally, has decided to take the school to the, until recently, inaccessible indigenous communities …
>
> (Instituto de Alfabetización en Lenguas Indígenas, 1945, pp. 140–160)

These integral strategies in education policy provoked an awareness through which the indigenist movement acquired prestige on a national level (Aguirre Beltrán, 1976). In 1940, Pazcuaro, Michoacán, became the elected headquarters for the first Encuentro Indigenista Interamericano (Inter-American Indigenist Summit) led by Moises Saenz—founder of the Departamento de Asuntos Indigenistas (Department of Indigenous Affairs, or DAI) and pioneer of the multiculturalist discourse. These activities, which fostered cooperation between education and research interests, were among many other new attempts to approach and understand indigenous cultures at the time. While such approaches transformed and indigenism took on new forms, this work remained, at its core, directly influenced by Gamio's work. For decades, politicians, educational and social reformers, and scholars in various fields shared the common belief that systematic studies of Mesoamerican civilizations, as well as contemporary indigenous society and culture, were vital to understanding and promoting México's unique cultural roots. The indigenists were progressive reformers and pioneers of their fields who sought to distance themselves from the ethnic homogenization pursued by the pre-revolution ruling class. Theirs is a mixed legacy, but one that has nevertheless laid a groundwork for all subsequent interventions and studies in Mexican indigenous communities.

Aesthetics and Design Practices in the Context of México

Before entering into the area of social practices in relation to design in México, it is important to first consider the transformation of readings of the popular arts, as this is unavoidably linked to the political context. "Popular arts' is a broad term referring to handcrafts, decorative arts, and folk arts—all of which are also encompassed by the Spanish word *artesanía*—and such arts share the common traits of being created by people to display the depth and variety of Mexican cultural heritage. Often, the simple definition given is that popular arts are the "the traditional, materialistic and formal expression of the common people' and those of indigenous heritage.

It is important to note that only after the first decade of the 20th century—in the context of revolutionary fervor and a newfound nationalism—was the first known detailed study of México's popular arts conducted (Florescano, 2006). In 1921, Jorge Enciso and Roberto Montenegro organized the first Popular Art Exhibition at Chapultepec Park in Mexico City. An unprecedented step toward México's nationalist modernity and aesthetic, with the approval of President Alvaro Obregón who saw this as an opportunity to craft a populist image for his regime. In 1922, several prominent artists led by painter Gerardo Murillo collaborated to compile the two-volume Folk Arts in México, which was published by the new government (López, 2010). Afterward, popular arts and their relation to concepts of identity became objects of discourse and political debate—seemingly parallel to the rise of Gamio's indeginist discourse. In the introduction to his 1930 volume of popular art Gabriel Fernández Ledesma wrote:

> En la civilización de las máquinas y el imperialismo comercial, rebaja la nobleza de la función mecánica invadiendo dominios de producción

> manual que no deberían desligarse nunca de su expresión de intimidad [...] viene la competencia a bajo precio, la disputa de mercados [...] el total de las industrias de México genuinamente son productos manuales [. . .] en los cuales confluyen una elaboración técnica perfeccionada, un sentido de belleza amalgamado al cuerpo mismo del producto, dejando satisfecha así una nueva necesidad común.
>
> (Fernández Ledesma, 1930, p. 1)

> The civilization of machines and commercial imperialism reduces the nobility of the mechanical function, invading the domain of manual production that should never be separated from its expression of intimacy [...] follows the competition of low prices, the fight over markets, [. . .] the totality of Mexican industries are genuinely manual products [...] in which a perfected technical production coalesce with a sense of beauty fused to the very body of the product, thus satisfying a new common need.
>
> (Fernández Ledesma, 1930, p. 1)

While compiling artifacts, Mexican intellectuals introduced a new discourse on the relationships between the aesthetic, imaginative force, and traditional myths. At the same time, educator and politician Adolf Best Maugard (1923), who led the Secretariado de Educación Pública (Public Education Department) from 1921 to 1924, published his book on *Métodos de Dibujo* (*Drawing Methods*), which emphasized the social function of art. The Best Maugard method, designed for use in México's art schools, provided a theoretical foundation for creating and analyzing art based on concepts of national identity and using symbols deemed universal. With the aim of students producing "purely Mexican" designs, Maugard based his primary exercises on seven graphic elements which were intended to create their own artistic language.

In the Boletín de Secretaria de Educación Pública (Bulletin of the Secretary of Public Education), Fernando Pontones wrote, "Consider only that those people that have genuine expressions in their artistic representations are the only ones that can be considered of the first importance in the human artistic evolution" (Best Pontones, 1922, pp. 227–229). In this same educational sphere, Saenz—initially sharing Gamio's assimilationist vision—worked to share knowledge of the markets with artisans. His goals were twofold: to establish commercial links that would support the growth of productive communities while also generating a national aesthetic. The dual influence of these goals guided the direction of cultural interventions in the 1920s and 1930s and the exaltation of popular arts, aesthetics, and Mexican cultural identity of the time (López, 2010).

Artisanal production after 1920, and during these first decades of independent México, had to confront two competing forces: first being the expansion of imports as a consequence of the country's open-doors strategy with Europe, and second, industrialization. We often consider crafts to be the precursor to what can be considered the "profession" of design, but historically, the practices of these crafts were relegated to the background and were overshadowed by the transformation of industrial processes. As a consequence, these creative practices and their associated skills—*el hacer* (the making)—have been separated as the common conception of the artisan is more associated with virtuosity than with the activities of design (Makovicky, 2011). Meanwhile, the artisans, usually creating in isolated areas—where, among other things, the artisanal structure coincides with geography, turning it into a sign of identity—ensured the continuity of Mesoamerican culture (Florescano, 2006).

When artisan's communal structures and traditional techniques disappear, a piece of their identity and cultural differentiation disappear too. This project aimed to work with communities that requested support for increased market exposure and, in some cases, whose artisanal techniques are on the verge of disappearing. This was a particular threat in the Santo Domingo Tonaltepec community, where the pre-Hispanic technique used with clay was maintained by only two artisans. While aiming to preserve these techniques, this project does not also aim to force the artisans to remain unchanged; rather, it provides a context so that artisans themselves may find a place for their techniques through material explorations within the context of the 21st century. With the project workshops, we facilitators did not intend to choose what to preserve or develop, but we sought to provide a space for an exchange of knowledges and the exploration of processes for artisans to transform what they deemed appropriate to achieve their goals.

As an example, at one of the workshops we held in Santo Domingo Tonaltepec, we invited an artisan from another town to share her technique of polishing clay with stone with the other local artisans. In this way, artisans in the community were exposed to a new traditional technique that they could test and would eventually apply in their own processes. This exposure allowed the artisans to decide if, within their process of "the making," the new method of polishing clay could provide them with any benefits to their own artifacts. Therefore, each artisan could decide whether or not to incorporate it into their own way of "making."

As I have previously noted, Mexican authorities—within the government and also politically connected organizations—have both worked to anchor and nullify popular art in their pursuit of constructing a national identity. In that sense, cultural artifacts have always been used as negotiating weapons; the value of each has been established according to local standards of authenticity and their ability to evoke national identity. The sheer fact that an artifact belonged to an indigenous community was seen as a key element of reaching a national identity (Thomas, 2002). With this understanding, Bonfil (1987) points out that official agencies used the discovery of archaeological sites and the commercialization of indigenous artifacts as strategies to stimulate tourism.

Experts in Latin-American culture, such as Roa Bastos and Juan Acha, point out that traditional methods are allowed to exist within their cultural context through adaptation which is often forced as the general mechanism of an imposed homogeneous system without hindering or questioning its development. Historically, hegemonic societies have tried to manipulate those practices that do not conform to the national homogenizing plan by sabotaging the transformation of any forms which question or alter a course set by the capitalist system. Some strategies include "folklorization", by which discourse is manipulated to distort the meanings of cultural artifacts in order to trivialize their symbols and their deep and spiritual content (Acha et al., 2004; Roa Bastos, 1980).[5]

The Anti-indigenist and Post-indigenist Critique

When Aguirre Beltrán entered into the field of Mexican indigenist policy, Gamio's voice and legacy remained virtually unchecked. The two met in 1942 when Aguirre Beltran was working as a physician in the town of Huatusco. Gamio encouraged him to publish his first book, a historical study of black people in México, in 1946. But over decades of indigenist work, Aguirre Beltrán departed from Gamio's tradition regarding some significant points of form and substance. From 1950 onward, there was serious criticism surrounding the traditional studies of community; it was obviously impossible to approach and understand a community, or ethnographically describe it in a sufficient manner, from the perspective of attempts toward a national unity. Realizing this, Aguirre Beltrán wanted to conduct his studies of communities in full view of their seemingly isolated social structure without the ambition to alter or intervene.

By 1956, Aguirre Beltrán was able to empirically establish the existence of a system of regions. He determined there was a center—conditioned by the presence of a ladino, or mestizo race—surrounded by a "constellation" of subordinate indigenous communities. Aguirre Beltrán's model, sometimes referred to as the "solar system", was formulated from the empirical evidence he collected in his research of the Tzeltal-Tzotzil region and other parts of México (Aguirre Beltrán, 1976). Two decades later in 1977, Ricardo Pozas, another prominent indigenist anthropologist, published a biography of a Tzotzil man named Juan Perez with a detailed work on the Chamulas heights of Chiapas' highlands. The organizational structures of these works were clear departures from Gamio's ideas.

In their 1954 volume titled *Métodos y Resultados de la Política Indigenista en México* (*Methods and Results of Indigenist Policy in México*), published by Alfonso Caso's INI, Pozas and Aguirre Beltrán described the concept of a community as mystically linked to its land. This unity, as they described it, had no conscience of being a fragment of some larger ethnic group, and feels even less than part of a nation, as the politics of the moment pretended. Their concept of community was entirely contained in the "nuclear" society that the indigenous person was bound to; thus, the concept of nation held no significance since it exceeded the indigenous person's sense of belonging. In this sense, Pozas and Aguirre Beltrán seemed to borrow from Caso's visions and theoretical framework while simultaneously critiquing the nation-building premise that he shared with other contemporaneous indigenists. In 1965, anthropologist Calixta Gutiérrez Holmes published an important follow-up to this new field of thought in her book *Los Peligros de las Almas* (*The Dangers of the Souls*), where she described the Tzotzil world of the Chiapas highlands.

Bonfil Batalla was the first in his field to systematically criticize indigenism and indigenist policy in his 1970 manifesto titled *De eso que Llaman la Antropología Mexicana* (*On that Thing Called Mexican Anthropology*). The book took issue with the actions taken until that point regarding field research and its relationship with the indigenous cultures and national identity (Warman, 1970). Bonfil's work, particularly *México Profundo* (*Deep México*) (1987) developed a discourse for the review and criticism of indigenist policy and anthropological practice in post-revolution México and is probably one of the most significant influences in my own inquiry. He argues that it is impossible to understand cultural features in terms of comparative values such as degrees of contemporary and traditional, or urban and rural, since these scales necessarily arise from the imposition of a Western perspective. Terms such as "primitive", "advanced", "modern", and "urban" do not correspond to true patterns of inner development.

Bonfil (1987) claims that cultural diversity is not a limiting factor to México's national identity. He argues that another force acts against México's ability to become a cultural capital although the possibility exists. Rather than focus on cultural diversity as a threat, Bonfil asserts that we need to return to the focal point of issues in Mexican society: colonization as the primordial problem. Bonfil (1987) established a distinct departure from the tenets of the indigenist policy and traced its lineal discourse (Warman, 1970). Meanwhile, he established an inflection point in relation to education, politics, and the hegemony of Europeanism. He was continuously encouraged by the 1968 student movement and later by anarchist Zapatism. Bonfil declared the need to evaluate different cultural expressions not along lines of simple comparisons and hierarchies of values—as had been the case before him—but through understanding each cultural expression in their own context.

Bonfil begins his review of indigenist policy by considering the latent national dichotomy set in motion by colonization. On the side of the indigenous culture, Bonfil defines "deep México" as the undercurrent of Mesoamerican civilization which refuses to accept outside imposition. He also establishes parameters for that resistance to imposed patterns along the lines of communal action to reject innovation or to appropriate the dominant cultural elements. These aspects of Bonfil's work contributed to my development of relevant parameters in this project's fieldwork and served to formulate a vision near the perspectives of the indigenous world and its realities.

Years later, Villoro (1998) proposed another revision of Gamio's work focused on his "forging of a nation" concept. Villoro argues that such an effort should not aim to assimilate distinct aspects of a nation into a homogenous and singular identity, but rather, should search for and/or develop a "multiplicity." Villoro proposes:

> Lo que se trata es de aceptar una realidad: la multiplicidad de las diversas culturas, de cuya relación autónoma nacería una unidad. Frente al Estado-nación homogéneo, se abre ahora la posibilidad de un Estado plural que se adecue a la realidad social, constituida por una multiplicidad de etnias, culturas, comunidades. Este Estado terminaría con el intento alocado de imponerle por la violencia un esquema pretendidamente racional. Tendría que ser un Estado respetuoso de todas las diferencias. Sería un Estado en que ningún pueblo, ni siquiera el mayoritario, impondría a otros su idea de nación.
>
> (Villoro, 1998)

> This is about accepting a reality: the multiplicity of different cultures, from whose autonomous relationship a unity would be born. Opposite the homogeneous Nation- State, the opportunity now arises for a plural State that fits the social reality, made up of a multiplicity of ethnicities, cultures, communities. This State would end the crazy attempt to impose through violence a supposedly rational scheme. It would have to be a State respectful of all differences. It would be a State where no people, not even the majority, would impose their idea of a nation onto others.
>
> (Villoro, 1998)

As Néstor García Canclini states: "las identidades tienden a esencializarse y la pertenencia comunitaria se ha vuelto la principal garantía de los derechos individuales [Identities tend to essentialize themselves, and belonging to a community has become the main assurance of individual rights]" (García Canclini, 2000, p. 109). Canclini's emphasis on "assurance of individual rights" and Villoro's mention of "the crazy attempt to impose violence' allude to the perils and exceptionally high costs of homogenization imposed upon marginalized communities by powerful agents and their respective organizations.

The works of these anti-indigenist, or post-indigenist, scholars present a radical critique of the methodology that shaped anthropological practice in México for decades. This literature highlights the fallacies of relying on distinctions between "advanced" and "primitive" cultures as such delineations provided the justifications for exploitation, erasure, and appropriation of cultures. Above all else, these works addressed the false narrative of harmony through nation building and formulated a powerful argument against a totalizing, and homogenizing, nationalist approach by positioning concepts of multiplicity and difference as critical and natural aspects of México's social reality.

Resilience Philosophy

Human beings have always forged communities with the desire to live and work together. These communal developments contribute to various sets of common beliefs embodied by cultural and political institutions. These structures tend to emerge organically to represent one of the fundamental features of human civilization: the communal expression of the marketplace (Papanek, 1995). Furthermore, the persistence of communal organizations in the economy and the cultures of indigenous peoples have always had a complex relationship with imposed homogeneous systems. This conflict, persisting throughout history, is due not only to the rampant penetration and destruction of existing indigenous cultures, but also due to their independent governing systems (Acha et al., 2004).

Although one the interest of this project was to understand local identity through the practice of social design, I will revisit the cultural transformation of the artisan role in México's indigenous communities. As I have previously mentioned, Jorge Pellicer's work titled *Los artesanos del porvenir* (*Artisans of the Future,* 1995) presents a case study of Santa Clara del Cobre. This project was coordinated and conducted in a public arts and craft school called CECATI 166, under the supervision of the American sculptor James Metcalf and Mexican artist Ana Pellicer. In their work as social design practitioners, Pellicer and Metcalf apply artisan processes as a way to support the resiliency and cultural heritage of the community[6]. Pellicer's and Metcalf's work provides considerable insight into how traditional production practices serve to bolster the strength of indigenous communities.

Communal practices to produce cultural artifacts contribute to the community's longevity and help preserve its culture. This maintenance of cultural continuity in indigenous communities has relied greatly on each community's own mechanisms for transmitting cultural knowledge across time and between generations. These mechanisms rely on cultural artifacts and traditions not only as a means to remember but also to enable the freedom to forget.

Several works on collective memory review and compare different communities' methods for associating knowledge of the past with the present so as to make decisions about the future. Jan Assmann, a celebrated Egyptologist, shows that the collective memory of spatial and temporal elements, and the communication methods specific to each community, coexist on an existential level. Stored within them are the ideas, emotions, and values that define the identity and culture of the community. He thus demonstrates how the collective memory not only rebuilds the past but also serves as a template to organize the experiences of the present and the future (Assmann, 2011). Similarly, Bonfil (1987) demonstrates how collective memory can filter the perception of new elements, thus contributing to decisions about the degree and manner to which they will be integrated into the community and cultural identity.

For many Latin-American indigenous cultures, and specifically those in México, the popular creative and transformative artifacts become most tangible during socio-religious festivities and the artistic expressions they include. These unique artifacts contain a collective memory of the creative features of the community and its cultural identity. The expressions contained within an artifact are usually collective since individual acts alone have had less of an effect over a shared history and the development of social codes. Even if the Western vision is able to identify with these artifacts on an aesthetic level, it cannot conceive of the cultural information contained deep within the social conditions and community ethos of the artifact (Acha et al., 2004).

Anthropologist Marta Turok's studies provide a benchmark within current works on Mexican craftsmanship and artisan production. Rather than categorizing artifacts based on materials and resultant forms, Turok (1988) describes the artisan's work as a reflection of production processes. She sets forth a vision for new systems of artisan qualifications and production itself. She claims that, "la producción artesanal de México enfrenta una nueva encrucijada ante los embates de la globalización [artisan production in México faces a new crossroads in light of the onslaught of globalization] (Turok, 1988)." In order to protect artisanal production processes and the quality of artisanal work, Turok suggests promoting products in traveling exhibitions and through local competitions.

Turok praises Victoria Novelo's (1976) work for her analyses of artifact production, including the instruments used and human relationships associated with craft practices. Important to this work, Turok and Novelo link the expressions of cultural knowledge embedded in el hacer (the making) with the social and economic spheres that Malinowski addressed years before. This link generates a common thread among forms of artisan production, productive composition of the community, and relationships between the local and national economy (Turok, 1988).

It is notable that for most of post-revolutionary Mexican history, the drive for a national narrative has entombed indigenous culture and traditions. México's nation building effort has relegated these cultures to the past or objectified them through a primitivistic lens. In the process, they are denied any possibility for change, and the possibility for transformation of their cultural artifacts is stymied. Acha writes:

> La idea de que lo popular, especialmente lo indígena es que debe permanecer idéntico a sí mismo, detenido en un punto anterior a su propia historia, está en el centro de uno de los mitos más características de la cultura occidental. [. . .] El artefacto popular está condenado a mantenerse originario y puro: si se transforma se pervierte, si incorpora novedades traiciona sus esencias, adultera sus verdaderos valores y corrompe su autenticidad primera.
>
> (Acha et al., 2004, p. 157)

> The idea that everything popular, especially everything indigenous, must remain identical to itself, arrested in a point previous to its own history, is at the center of one of the most characteristic myths of the western culture [. . .] Popular artefacts are doomed to remain original and pure: if it is transformed it is corrupted, if it incorporates novelties, it betrays its essence, adulterates its true values and corrupts its first authenticity.
>
> (Acha et al., 2004, p. 157)

Indigenous communities have their own processes—natural conditions for the transformation of the artifact and its mode of cultural expression—which answer only to the communities' need for continuity. The appropriation of foreign elements and processes answers an internal need to adapt and is introduced according to the agreed-upon needs of a group with communal cultural affinities.

As moderate as these foreign involvements might seem, they can still disrupt and distort the indigenous cultural processes and their meanings:

> Visto desde fuera un cuerpo/artefacto cultural tiene una exagerada fragilidad. Una pequeña presión es suficiente para lesionarla. Pero considerándolo desde dentro es vigoroso y resistente: puede incorporar grandes pesos y soportar bruscas sacudidas sin comprometerse su derrotero.
>
> (Acha, 1994, p. 235)

> Seen from outside, a cultural body/artefact has an exaggerated fragility. A little pressure is enough to harm it. But considering it from the inside, it is vigorous and resistant: it can incorporate huge weights and bear sudden shakes without compromising its course.
>
> (Acha, 1994, p. 235)

Communities should reserve the right to decide which elements they will maintain, incorporate, or substitute in their traditional practices. As shocking as this acculturation process may seem to outsiders, it resolves itself naturally and can produce well-integrated cultural expressions.

According to sociologist Caroline Knowles, "Resilience is about people's ability to adapt to conditions generated by different tempos of social change, conditions which create ever-new forms of social fragility" (Knowles, 2014). Similarly, Tunstall suggests the need to accept a community's value system as something dynamic, not static; each successive generation must be allowed to negotiate for themselves the elements that constitute their values and culture. This means eliminating the labels that draw distinctions between "popular arts" and "design" and instead focusing on what she calls "forms of making" as the method by which communities make their value systems tangible to themselves and others (Tunstall, 2013). The acculturation process is fundamental to exchanges among cultures and to the transformation of communities for their continuity.

The approach throughout this project draws directly from the works that I have now discussed and ultimately transformed through experimentations and discussions with the project's other design facilitators and artisans over the course of our collaborative work. We facilitators considered cultural artifacts with regard to their intangible value as a carrier and transmitter of the cultural elements of each community. This perspective also aimed to avoid "preservation" from the western point of view, so as to avoid denying the community possibilities for change. We also viewed traditional techniques as dynamic "forms of making" as Tunstall describes them.

In this present work, we facilitators do not consider the making—as it is referred to by Turok and Novelo—as something immovable and untouchable. Nor was it our mission to decide or select which parts were to be preserved or transformed, or to approach it as a process of linear development. We view these perspectives to be a limited aim toward a specific goal, rather than accept the goals and needs of indigenous communities as they see them. We believed that it was up to members of each community to define their own goals toward and models of resilience against processes of homogenization. If the communities felt that it was appropriate, they were able to adopt tools and techniques that we all shared during the workshops.

We facilitators did not engage in active decision-making set out unilaterally to achieve some pre-established agenda. Instead, decisions were made jointly through a social design practice and relied on a democratic process including all participants of the workshops—facilitators and artisans alike. Our mission was to observe and reflect upon our experiences in the workshops in order to create environments where artisans could decide for themselves whether or not they wanted to adopt the tools and methods that we could exchange.

Expressions of Exchange

As I have previously noted earlier, scholars who study indigenous cultures and the importance of craftsmanship tend to link the transmission of traditional cultural knowledge and the artisanal production process. For any group existing in the periphery of the industrialized world—from Middle Age workshops to indigenous craftsmanship—the community serves as the context and medium in which to learn; where both aesthetic judgment and coherent style are developed according to the common values of a cultural group.

In this way, cultural artifacts are both the physical and the symbolic result of a community's creative expressions. But while such artifacts have traditionally been linked to rural forms of self-sustainment and trade—where use-values prevail over exchange-values—communities now seem to be producing their artifacts for sale rather than for their own use. Artifacts and the right to their possession now circulates in a broader economic system through the exchange of money, and at that precise moment—as Alfred Gell (1998) points out—the conceptual commodification of the artifact takes place. Production and consumption patterns are the basic elements of any market system; cultures vary in how these exchanges are structured and how they relate to the social membrane of the community (Gell, 1998).

Tracing back the origins of artifacts' reveals many of the inner-workings of the social structures they come from. This key insight seemed to drive Boas' detailed study on México's local languages as well as Malinowski's studies on the market in relation to communities, both of which focused specifically on Oaxacan communities. Boas and Malinowski both occasionally used a method to observe many people interacting—often in hectic marketplaces—for long periods of time in order to decipher a social structure that went beyond mere coincidences. This allowed them to plan their studies and to generate an approach to a community and its ways of exchange (Murphy and Marcus, 2013; Malinowski, 1957).

Malinowski discovered a strange but unmistakable relationship between spatial and temporal elements and market fluxes. He came to understand that the markets were as much about sociocultural factors as they were about the tastes and preferences of the population in and around the Oaxaca Valley (Malinowski, 1957). Malinowski was able to differentiate between traders, producers, and consumers. In doing so, he was also able to differentiate between types of artifacts by separating them into primary production, native production, local crafts, and imported goods. This allowed Malinowski to focus on the corn market fluxes as corn set the price levels for other important goods such as meat, bread, vegetables and clay utensils (Malinowski, 1957).

While the modernization myth tends to equate indigenous societies with a primitive notion of pre-capitalism, it seems that many indigenous societies conform to an alternative socio-political system altogether. It may also not be accurate to confine indigenous societies to the label "pre-capitalist" if they are actually "a-capitalist." These narratives, dating back to Adams Smith, continue to paint indigenous societies as backward although they may choose to engage with the global capitalist market or remain a-capitalist and apart from such a system by choice. The latter decision should not imply that a community is backward or "pre-capitalist"; instead, it goes to show how local indigenous communities hold alternative views about our shared world. It must be noted that these societies conform to different systems of beliefs that may make little sense when viewed through the reductive lens of Western values, history, and identity politics.

This present work does not view artifacts as locked traditions but rather as cultural expressions that are living, transforming, and adapting. Under these parameters, artifacts' values may be transmitted through exchange once an environment allowing the sufficient space and time is created for the development of new social relationships and structures. Exchanges that take place under this new sort of arrangement should allow for new forms of expression to take shape. This is a necessary collective process, as the exchange of values only occurs through a cultural correspondence. This living condition of cultural knowledge and memory is capable of profound social restructuring to influence the economic and cultural characteristics of a society (Kopytoff, 1986; Pellicer, 1995; Kester, 2011).

Design Before Design: Approaching People, Objects, and Outcomes

A fundamental aspect to this project was our facilitator's humanistic approach to the subjects and cultural artifacts of indigenous communities. Building off Igor Kopytoff and others who discussed objects as operating within a circulation of meaning, the project suggests other avenues for building close ties with indigenous communities. During the project, we facilitators approached the complex processes of cultural continuity and adjustments to change as elements of resilience and identity. We connected threads throughout the cultural materials of the community, which allowed all participants to understand and reflect upon socio-historical transformations.

Caroline Knowles (2014), in her ethnographic portrait of "globalizations backroads,' argues that a characterization of the movement of people and things as a "flow" is erroneous and suggests that we might rather call it a "journey." By following the journey of the artifact—from the production and distribution lines where they begin their movement—I endeavor to uncover and articulate the human and social relationships that develop along the way. In my reflection, I perceive social design as a practice for creating a narrative thread. This approach leads to my dual understanding of the social design practice in this scenario: on the one hand, it is an act of creating spaces for resistance and self-definition; and on the other hand, it is a communal journey for cultural continuity and transformation through means of the artifact.

The anthropological study of objects can give shape to an exploration of culture and opens many lines of inquiry when we realize the ways in which they intertwine with peoples' lived experiences (Malinowsky, 1957; Knowles, 2014). When we understand design as an ancestral activity occurring over the entire course of human history rather than a more recent occupation, we can begin to change perceptions about the content of indigenous, African, Asian, or Latin-American designs. We are able to ask new questions about the issues of design as we expand our understandings of design to include new forms.

This work emerges from a long discourse which questions the appreciation and consumption of culture from a Western capitalist perspective. Victor Papanek was among the first to criticize "design conceptualization" in modern society. Papanek and Gui Bonsiepe (2012) argued over the role of design in "peripheral" countries—as Bonsiepe called them—but they were united in their concern for finding a responsible use of design. Throughout his career, Papanek desired to understand design from the perspective of indigenous people and wove his interest into the designs through vernacular forms linked to places, materials, and ways of production. In Design for the Real World, Papanek criticized the profession for its affiliation with the concept of capitalist well-being—in its emptiness and mediocrity—and the lack of authenticity in Western society's objects (Clarke, 2011; Papanek, 1971). This critique of industrial society bore a striking resemblance to the vision which formed the basis for William Morris's arts and crafts movement in Great Britain 100 years earlier[7]. But anthropology had already begun to see the possibilities of artifacts and their ability to transgress, to self-modify, to stir the culture, and to also cling to it. As anthropologist Nicolette Makovicky (2011) notes, the best artifacts have always been sublimated to mere harbingers of modern design, thus eliminating any other potential within the economy and freezing them in place, denying them the possibility for transformation. This generated a schism in the craft-making process, separating creativity from actual skills of craftsmanship which, as mentioned, was forever linked to what Alison J. Clarke (2011) calls "manual virtuosity." Thus, a long line of thinkers criticized the tradition of judging an artifact's design from the perspective of Western markets and tastes. Through this lens, the culture of the object becomes a redundant factor, and the value of the process of "*el hacer*"—"the making"—is lost. The training, the skill, and the materials involved in craft-making should be understood as a dynamic practice of correspondence.

The reach of the Western canon leaves creators of cultural artifacts in indigenous communities relegated to a second-class position within the world of high art. This has been a historical constant since even before the American conquest, affecting other expressions such as pottery and metalwork. This loss of appreciation for the indigenous cultural artifact has limited it, nearly without exception, to the fields of anthropology and archaeology rather than fine arts (Acha et al., 2004).

Salvador Toscano, considered México's first researcher on Mesoamerican aesthetics, was interested in Mesoamerican art and expressions throughout his lifetime. His research often aimed to configure an indigenous aesthetic. He adds an important critical view on the Western position, writing:

> El estilo artístico es la fisonomía, la forma por la cual se expresa una cultura, su expresión psicológica peculiar; no existe, por lo mismo, un criterio de validez universal que nos permita juzgar el arte de los diversos pueblos en su desarrollo histórico, pues ni siquiera el ideal clásico de los griegos –tradicionalmente señalado como el momento más alto de la humanidad en el arte– puede reclamar tal título. No existen artes bárbaras e inferiores, pues los estilos artísticos no son mejores ni peores, sino diferentes.
>
> (Toscano, 1970, p. 13)

> The artistic style is the physiognomy, the way a culture expresses itself, its unique psychological expression; there is not, for this reason, a universally-valid criterion that allows us to judge the art of the various peoples in their historic development, since not even the classical ideal of the Greeks—traditionally agreed-upon as humanity's pinnacle in art—can claim that title. There are no barbaric and inferior arts, for artistic styles are not better or worse, but only different.
>
> (Toscano, 1970, p. 13)

In a similar vein, Acha points out that in Western culture, artifacts largely remained attached to natural forms until the end of the 19th century while indigenous artifacts have remained, since the beginning, viewed with a tendency toward abstraction (Acha et al., 2004).

Significantly, Geertz (1983) identifies "the system of modern culture" as a machine through which objects collected from non-Western cultures are classified in two main categories: the scientific, which terms them "artifacts";

Figure 3.1 Mixtec textile samples. Iconography representing the landscape of the Mixtec highlands. CADA Foundation®, 2015.

and the aesthetic, where they become "works of art." However, as Gell (1998) points out, if Western aesthetic theory is supposed to apply to "our" art, then it must be equally applied to all. He writes that, "It makes no sense to develop a "theory of art" for our work and a different one for those cultures that fell under the yoke of colonialism" (Gell, 1998). From the perspective of contemporary Western society, it is important to note that Acha warns how aestheticism can be isolated from the different factors involved in artifact production (Acha et al., 2004).

Expanding upon Geertz's (1987) theory of division within the "system of modern culture," Sofaer (2007) argues that historically, there have been two approaches to materiality. On the one hand, the prevailing art history discourse emphasizes the ways in which "objects provoke aesthetic responses," and artisans and artists are perceived through their manipulation of materials. On the other hand, is the discourse of material culture emphasizing the relationship between person and object, and considering the artifact's influence in communal life both locally and globally "by encoding or producing meanings"—both aesthetic and non-aesthetic responses—that create or challenge the values attached to human relations (Sofaer, 2007). This second discourse and approach to materiality is where this project aligns.

In a 2002 volume compiled by Fred Myers, Christopher Steiner reviews the evolving theoretical arguments under this line of thought. Steiner reflects on Bourdieu's 1977 foundational text for anthropology and sociology, *Outline of a Theory of Practice*, in which he discusses the object's ability to imply human conditions. He also reviews Hegel's work in his 1807 *Phenomenology of Spirit*, in which he asserts that everything we do and are reflects the process by which we also generate material forms. Bruno Latour (2005) and Gell (1998) add a further layer to the argument about an object's ability to generate "agency"—an attribute previously and generally associated only to individuals. In doing so, they open a new discourse about how artifacts reveal complex intentionalities. Gell (1998) goes even further by exploring the psychology of patterns, of control over knowledge, and of the interpretation of meanings. Steiner (2002), however, opposed this reasoning. He writes:

> I have already suggested, a hasty reading of Kopytoff's model of the "cultural biography of things"—which has become the dominant model for interpreting the movement of material objects through channels of trade and exchange—gives too much authority to objects and not enough to those who imbue objects with meaning. By shifting our focus from the "disembodiment" of objects in motion to their embodiment in the sociocultural space of legal and judicial process, we can begin to see material culture once again as "objects", without claims of subjectivity or personhood, whose meanings are opportunistically readjusted within the legislative framework of classificatory bureaucracies.
>
> (Steiner, in Myers, 2002, p. 211)

Political theorist Jane Bennett, known for her work on nature and ethics in the environment, also credited the object with an ability to generate agency by claiming that "political theory needs to do a better job of recognizing the active participation of nonhuman forces in events" (Bennett, 2010). Furthermore, according to Mexican anthropologist and educator Fernando Martín Juez, the object has several life stages. He writes,

> [. . .] emociones cotidianas y signos comunes se adhieren a las superficies de los diseños, los embozan con aquellas metáforas que más convienen al sistema de creencias de las comunidades a las que pertenecen.
>
> (Martín Juez, 2002)

> [. . .] everyday emotions and common signs adhere to the surfaces of the designs; they wrap them with those metaphors which are more convenient to the belief system of the communities they belong to.
>
> (Martín Juez, 2002)

Artifacts are thus considered important sources of information to analyze not only for what they say about their contexts but also for the ways that people are affected by their attributes. These are the ideas echoed in Gilda Hernández's work on pottery (Hernández Sánchez, 2013), which exemplifies the post-structuralist work which often attributes subjects' qualities to objects.

Anthropologist Paul Connerton claims that our knowledge of all previous human activity is only possible through the material traces that they have left us. Historical reconstruction, therefore, does not depend on social memory. On this topic, Bennett (2010) reviews Spinoza's work and says that he "ascribes to bodies a peculiar vitality, as every nonhuman body shares with every human body a conative nature". Thus, artifacts allow for the exploration of how members of our society act, and in deciphering the way they are produced, we can reveal how their relations operate in many different contexts (Connerton, 1989).

Besides the various attributes reviewed above, it is important to note that artifacts have an important time dimension which allows us to document their past (Hernández Sánchez, 2013; Knowles, 2014). This time dimension follows a more conservative tendency toward pattern repetition in Latin-America which is more influenced by community dynamics than in other cultural systems (Acha et al., 2004). About artifacts, Baharam writes, "they acquire meanings in use, become integrated in everyone's whole life experiences, and interact with the mythology from which they derive their symbolic strength" (Balaram, 1995, p. 143). Capitalism, however, has interrupted the virtuous cycle by which we can create the objects that express an understanding of who we are or can be. This constitutes a loss of humanism at the hands of commodification and consumer brands. As such, the artifacts transformed during this project are considered to have cultural properties by their respective communities in relation to their roles as community witnesses. The artifact production will play a significant part in the social life of these artisan communities who seek greater exposure to new market systems.

A Humanistic Approach

I have outlined the social and political factors that have shaped the indigenous narrative and policy in México by looking more specifically into how indigenous culture has been used as part of the nationalistic narrative during the post-revolution period. I have also taken a humanistic approach in response to the capitalist economic framework, which takes important cultural artifacts to a point where they completely lose their socio-historical meaning. By uncovering the human relationships that occur in the production and transmission of cultural artifacts, cultural materials can be viewed as a reflection of the social and cultural transformations a community may make.

With reference to Gamio's holistic analysis of indigenous populations in the Teotihuacan Valley, I have discussed how powerful agents and institutions have used indigenous peoples and cultures as instruments in a post-revolutionary nation-building effort. These nationalist narratives and the indigenist policies of post-revolution México contained a broader goal of assimilation and sought to integrate indigenous communities into mainstream society by force. While this assimilationist policy distorted the cultural meaning of objects to fit a specific nationalist agenda, it leaves a mixed legacy because of the significant groundwork it laid for subsequent research in this area. Post-indigenism evolved from previous assimilationist efforts and sought to conduct studies on indigenous populations within the context of their isolated social structure without any ambition to intervene or influence the community. This approach attempts to view cultural expressions within their own context and moves away from analyzing indigenous communities through a Western framework which tends to reduce the importance of tradition through the lenses of modernity and linear progression.

I have also discussed this project's humanistic approach to people and objects by emphasizing the significant social role that cultural artifacts play in indigenous communities. While it is important to acknowledge the risk of homogenization and forced assimilation when studying interventionist approaches to indigenous communities, it is equally important not to romanticize indigenous cultures by viewing them as fixed traditions that are immune to—or ought to be protected from—transformation. Foreign interventions can address the internal needs of indigenous communities as long as these interventions are agreed upon by the group and equally provide a platform for the transmission of values. Through a humanistic approach with an emphasis on equal collaboration and correspondence, it may be possible to achieve cultural continuity in indigenous communities that seek the opportunity to transform their own practices and achieve their own goals. From a review of previous studies and existing theoretical research, it is notable how many complex historical, cultural, and social factors have shaped and shifted the realities of artisan's ecosystems without much reference to their needs and goals.

I hope that my discussion of circumstances that have shaped the cultural context of artisanship will offer some perspective into my own positioning as I formulate the processes of this project's workshops. It is important to note the fundamental contextual background that the facilitators and myself have gained in order to review our grasp, understanding, and navigation of cultural differences during the workshops. Over the course of the project, we utilized this background as a reference during our observations and deep reflections of the many conversations, meetings, and interviews between all and over the span of years visiting México.

Notes

1 Translator note: The word "íntegro" has two main meanings in Spanish. The first means "complete" or "full" and alludes to the totality of something. The second meaning refers to people as "honest." The title of the book encompasses both meanings.

2 Those documents, and especially the visual supplement, are relevant because of their direct relation to my own fieldwork. In my mind, the functionalist methods in no way contradict a legitimate historical focus, as for Malinowski both analyses are possible in the same analytical context.

3 As the first, and one of the few, anthropologists to approach Oaxacan culture in a descriptive and reflective way, Malinowski's work—especially the images he and

De la Fuente captured—was incredibly influential on my own. Only after Brown's edition in 1982 was the extent of his immersion in the Oaxacan market system appreciated or understood, and still to this day it receives little attention.

4 In Latin American contexts, both colloquial and written, and as I´ve pointed out before, "Indian" is used as a synonym to "indigenous," or "native." Columbus, upon arrival in continental America, believed that he had landed in the Indies, which was eventually named the "West Indies"—Indias Occidentales in opposition to "oriental Indies" the Indian Subcontinent, land of richness and spices. Amerigo Vespucci would eventually lend his name to this new land: América.

5 See also the documentary film in Oaxaca. Artist: Vargas de la Maza, 1950. UNAM Universidad Nacional Autónoma de México, Filmoteca.

6 Centros de Capacitación para el Trabajo Industrial or CECATI are industrial Work Training Centers, part of the Directorate General for Work Training Centers (DGCFT), part of the Administrative Unit attached to the Under-secretariat of Higher Secondary Education (SEMS) of the Mexican Ministry of Public Education (SEP). Their aim is to provide Mexicans with a wide range of technical work opportunities, to improve their outputs, their quality of life, and aid peoples' economic and social development. CECATI offers over 200 courses which are grouped into 52 specialties based on the different productive and service branches of the country. Most of the courses do not require an academic background and offer a valuable opportunity for communities with limited resources and/or which are affected by geographic isolation.

7 See again, the discussion of William Morris and the social dimensions of the Arts and Crafts Movement in Chapter 2.

References

Acha, J. 1994, 2008. Las culturas Estéticas de América Latina. México DF: Trillas.

Acha, J., Colombrea, A. and Escober, T. 2004. Hacia una Teoría Americana del Arte. Buenos Aires: Ediciones del Sol.[Aguirre Beltrán, G. and Pozas Arciniega, R. 1973. Política Indigenista en México, La. Métodos y Resultados. Instituto Nacional Indigenista y Secretaría de Educación, México, DF.

Aguirre Beltrán, G. 1976. Obra Polémica. Tlalpan, D.F.: Instituto Nacional de Antropología e Historia.

Anderson, B. 1983. *Imagined communities*. London, New York: Verso.

Assmann, J. 2011. *Cultural memory and early civilization: writing, remembrance, and political imagination*. New York: Cambridge University Press.

Balaram, S. 1995. Product symbolism of Gandhi and its connection with Indian mythology. In: Margolin, V. and Buchanan, R. eds. *The idea of design*. Cambridge (Massachusetts), London: The MIT Press, pp. 127–143.

Bennett, J. 2010. *Vibrant matter: a political ecology of things*. Durham: Duke University Press.

Best Maugard, A. 1923. A new method for developing creative imagination. *University High School Journal*. 3(4), pp. 247–255.

Best Pontones, F. 1922. El Método Best Maugard para la enseñanza del dibujo y su aplicación a los trabajos manuales. Boletín de la Secretaría de Educación Pública (Mexico City). 1(2), pp. 227–229.

Bhabha, H. K. 1994. *The location of culture*. London and New York: Routledge Classics.

Boas, F. 1927. *Primitive art*. New York, Dover Publications, Inc.

Bonfil, G. 1987. México Profundo. Una civilización negada. México, DF.: Grijalbo
Bonsiepe, G. 2012. Diseño y Crisis. Valencia: Campgràfic Editors.
Bourdieu, P. 1977. *Outline of a theory of practice*. Cambridge University Press.
Caso, A. 1953. El Pueblo del Sol. México, D.F.: Fondo de Cultura Económica. Lecturas Mexicanas Ediciones 1983, pp. 145–181.
Clarke, A. J. 2011. *Design anthropology: object culture in the 21st century*. Viena: Springer.
Connerton, P. 1989. *How societies remember*. Cambridge: Cambridge University Press.
De La Fuente, J. 1964. Educación Antropológica y Desarrollo de la Comunidad. México DF.: Colección de Antropología Social. Instituto Nacional Indigenista.
Fanon, F. 1967. *Black skin, white masks*. New York: Grove Press.
Fernández Ledesma, G. 1930. Juguetes Mexicanos. México, D.F.: Talleres Gráficos de la Nación.
Florescano, E. 2006. El Juguete Mexicano. México D.F: Taurus.
Gamio, M. 1916. Forjando Patria. Michigan: University of Michigan Library.
García Canclini, N. 2000. La globalización imaginada. México: Paidós.
Gell, A. 1998. *Art and agency: an anthropological theory*. Oxford: Clarendon Press.
Hernández Sánchez, G. 2013. *Ceramics and the Spanish conquest*. Leiden: Brill.
Instituto de Alfabetización en Lenguas Indígenas. 1945. INEA. México.
Kester, G. H. 2011. *The one and the many, contemporary collaborative art in the global context*. London: Duke University Press.
Knowles, C. 2014. *Flip-flop: a journey through globalisation's backroads*. London: Pluto Press.
Kopytoff, I. 1986. The cultural biography of things: commoditization as process. In: Appadurai, A. ed. *The social life of things, commodities in cultural perspective*. 11th ed. Cambridge: Cambridge University Press, pp. 64–91.
Latour, B. 2005. *Reassembling the social: An introduction to actor-network-theory*. Oxford: Oxford University Press.
López, R. A. 2010. *Crafting México: Intellectuals, artisans and the state after the revolution*. Durham and London: Duke University Press.
Makovicky, N. 2011. Erotic needlework: vernacular designs on the 21st century market. In: Clarke, A.J. ed. *Design anthropology: object culture in the 21st century*. Viena: Springer, pp. 155–168.
Malinowski, B. and De La Fuente, J. 1957. La Economía de un Sistema de Mercados en México. In: Acta Anthropologica, Época 2, 1(2). México D.F.: Escuela Nacional de Antropología e Historia, Sociedad de Alumnos. (Drucker-Brown, S. 1982. Malinowski in México: the economics of a Mexican market system. London, Boston, Melbourne and Henley: Routledge & Kegan Paul.
Martín Juez, F. 2002. Contribuciones para una Antropología del Diseño. Barcelona: Editorial Gedisa.
Murphy, K. M. and Marcus, G. E. 2013. Epilogue. In: Gunn, W., Otto, T. and Smith, R. C. eds. *Design anthropology: theory and practice*. London and New York: Bloomsbury, pp. 251–268.
Myers, F. R. 2002. *The empire of things: regimes of value and material culture*. Santa Fe, NM and Oxford: School of American Research Press.
Papanek, V. 1971. *Design for the real world: human ecology and social change*. 2nd ed. Chicago: Academy Chicago Publishers.
Papanek, V. 1995. The future isn't what it used to be. In: Margolin, V. and Buchanan, R. eds. *The idea of design*. Cambridge, London: The MIT Press, pp. 56–69.
Pellicer, J. 1995. Artesanos del Porvenir. México, D.F.: Secretaría de Educación Pública, Universidad Autónoma Metropolitana.

Roa Bastos, A. 1980. Las culturas condenadas. 2nd ed. México: Siglo XXI.

Sáenz, M. 1939. México Íntegro. México, D.F.: Consejo Nacional para la Cultura y las Artes (1st ed. on CNA, 2007).

Sofaer, J. 2007. *Material identities*. Malden (USA), Oxford (UK) and Victoria (Australia): Blackwell Publishing, pp. 1–9.

Steiner, C. B. 2002. Rights of passage: on the liminal identity of art in the border zone. In: Myers, F. R. ed. *The empire of things: regimes of value and material culture*. Santa Fe, NM and Oxford: School of American Research Press.

Thomas, N. 2002. Appropriation/appreciation: settler modernism in Australia and New Zealand. In: Myers, F. R. ed. *The empire of things: regimes of value and material culture*. Santa Fe, NM and Oxford: School of American Research Press.

Toscano, S. 1970. Arte precolombino de México y de la América central. 3rd ed. México D.F.: Universidad Nacional Autónoma de México.

Tunstall, E. 2013. Decolonizing design innovation: design anthropology, critical anthropology, and indigenous knowledge. In: Gunn, W., Otto, T. and Smith, R. C. eds. Design anthropology: theory and practice. London and New York: Bloomsbury, pp. 232–250.

Turok, M. 1988. Cómo acercarse a la artesanía. México D.F.: Editorial Plaza y Valdés.

Villoro, L. 1998. Estado plural, pluralidad de culturas. México D.F.: Editorial Paidós/ Universidad Nacional Autónoma de México.

Warman, A. 1970. De eso que llaman la antropología Mexicana. México, D.F.: Editorial Nuestro Tiempo.

Part 2

The Journey

4 Design as a Practice of Correspondence

El Hacer—The Making

This chapter tells the background of our five-year project endeavor, starting with our aims to develop a collaborative methodology, and detailing the ways we deployed the methodology through workshops with the ultimate goal of designing new processes of co-creation between design practitioners and artisans. In this chapter, I also discuss how the project's facilitation team came together to work alongside artisans and transform our collaborative methodology across the span of the project's workshops. The first section will explain the aims and results of ethnographic fieldwork that led to synthesizing a comprehensive facilitator team representing a diversity of backgrounds and community connections. Additional context will be explained through community visits and interviews with relevant figures in the fields of Cooperation, Anthropology, and formal Craftsmanship training, along with detailing how we facilitators connected with participating artisan communities. The second section will elaborate on how the proximities of design methodology was developed through fieldwork and research as well as through reflective practice and observation that would organically transform the workshops by processes of making and unmaking.

In February 2013, I visited five towns in the vicinity of Oaxaca de Juárez to conduct ethnographic research and meet with communities who worked with a wide variety of mediums including red clay, black clay, wood, and textiles as well as weaving, pleating, and embroidering cotton garments. I also visited two communities who worked independently in the processing and manufacturing of silk. During this ethnographic research phase, I gathered a group of design practitioners who collaborated as facilitations and documentarists at different points during the process, and including a series of ongoing meetings to design participatory workshops. I used these meetings to share my theoretical background research, contrast my findings against their insights, and gain local perspectives on training and co-design. As the workshops proceeded, we pivoted the focus of our work toward developing longer term projects with two Mixtec communities, San Pablo Tijaltepec and Santo Domingo Tonaltepec, which resulted in the creation of textile and ceramics artifacts, respectively. We encouraged artisans to experiment with their traditional techniques within new contexts, such as a change in dimension from 2D to 3D objects—from

DOI: 10.4324/9781003360803-6

embroidered blouses to standing toys in the case of San Pablo Tijaltepec—or a change of scale—from large-scale functional pottery to small-scale decorative items using pre-Hispanic techniques in Santo Domingo Tonaltepec. The design team also exchanged alternative materials and/or tools with the artisans. In all cases, participating artisans had requested the services that we could provide as a team of designers. It remained critical to the integrity of the project that the artisans possessed the freedom to choose, reject, adopt, adapt, or transform their techniques based on our open exchange of knowledges. In essence, we facilitators aimed to support the artisans so that they could bolster individual and community resilience and agency against the encroaching forces of globalism.

The story told in this chapter articulates a wider comprehension of the cultural, geographical, and idiosyncratic aspects of the project environment, which affected the development of fieldwork. It provides a frame for Chapter 5, where the ethnographic research and two main case studies, development of workshops, and outcomes thus obtained are documented in precise detail. I believe that this detailed project case study will assist and inform other design research practitioners, scholars, and students in formulating a design methodology where they, serving as facilitators, can collaborate with artisans in other cultural co-creative contexts as long as they also consider specific socio-historical contexts in each new endeavor.

Developing a Design Process: Oral Tradition and Craft as Cultural Symbolism

It is important that I begin by contextualizing the role of Oral tradition and Craft as cultural symbolisms in cultures where a written language is not the predominant tool for knowledge transmission due to devastating consequences throughout history. During the workshops, our collective understanding of the role of oral tradition and its relation to art, craft, and aesthetics was a necessary first step when connecting with Oaxacan artisan communities. Oral narratives allowed us facilitators to better understand artisan traditions, mechanisms for knowledge transmission, meanings, and cultural values in both the past and present of different community contexts.

Oral and artisan craft traditions have been, for centuries, a way to transmit and keep alive indigenous peoples' cultures, values, cosmologies, and belief systems. They have generated orientation patterns for possible conflict resolution, creating symbolic systems to provide order and meaning to communities who practice them. Oral tradition and cultural artifacts are a manifestation of aesthetics, ethos, and ethics, and it would prove to be of particular importance in the context of Mexican cultural history, both indigenous and non-indigenous.

Within Mexican indigenous traditions, artifacts and production techniques, stories, songs, and legends are passed down from elders to younger people as a part of teaching and learning as well as to perpetuate community traditions from one generation to the next. Oral traditions are fundamental to major pre-Columbian Mexican and Mesoamerican cultures, such as Olmec, Mayan, Mixtec, Zapotec, and Aztec. The language alphabet consists of glyphs and drawn symbols instead of letters corresponding to syllables and words. Thus, graphism and artifacts, where represented, are transformed beyond their pictorial and utilitarian qualities, turning them into a unique culturally contextual legacy rich in symbolism, like "books" and differentiating cultural and ethnic nuances among communities. To quote Gossen and Ámbar Past on this point:

> Oral tradition should be considered as part of the "ethnography of speaking". Oral tradition is important for the maintenance of the social order as are kinship systems and social organization.
>
> (Gossen, 1974)

> ... a wise person is said to have: "books in the heart" according to Robert M. Laughlin's translation of a sixteenth-century Spanish – Tzotzil dictionary. Incantations. Song, Spells, and Images by Mayan Women.
>
> (Past, 2005)

In two detailed studies—on the approach to communities and their oral traditions as seen in Chamulas in the world of the sun titled Time and Space in a Mayan Oral Tradition (Gossen, 1974) and The Interpretation of Cultures (Geertz C., 1973)—both Gossen and Clifford Geertz are quoted as saying, "The community is ruled by its ethos which entails the ethics and aesthetics as integral elements of the community." Geertz's remarks serve as a core articulating concept throughout this research project because the community is indivisibly linked to the artifact's design through its uses and behavior which operate as emblematic elements of the community's own personality and culture. This explains how ethics and aesthetics are central to these communities' traditions since symbols, graphism, and artifacts are, in their own terms, documental elements for transmitting cultural legacy.

In approaching communities generally—and artisans particularly—by learning about and using their cultural symbolisms, aesthetics, and images, and by presenting other cultural references through accessible and known objects, we, as facilitators, aimed to build mutual trust and respect in order to generate constructive design collaboration during the project workshops. By understanding oral tradition as an intangible artifact, we developed a process by which we could hear and see individual artisan oral accounts from their collective traditional narratives that they then interpreted into material expressions. This process allowed the artisans to draw out personal insights—either interpretative or imaginative—that could be translated into tangible forms such as drawings or sculptural, 3D figures, which could then be shared and contrasted against the group's recollections and imagery.

In order to establish collaborative dialogue between facilitators and artisans, it was important to establish definitions of ethics and aesthetics, which refer to the sets of rules and stylistic elements characteristic of a group. The team and

I understood the workshops as spaces where the artisans could share their histories and cultures through spoken word, craft techniques, and explorations. We also viewed our role during the workshops as providing the necessary support and services initially sought by the artisans to remain resilient against global forces and in order to preserve their individual and communal agency. It was crucial to the development and transformation of these workshops that we established a set of flexible parameters within which facilitators and artisans could communicate across cultural references. Ultimately, the workshop process aimed to provide space for the different communities who sought and participated in our services to reflect upon and create artifacts inspired by their respective ethos.

The overall aim of the workshop program was to develop spaces for facilitating collaborative design through processes where design practitioners, acting as facilitators, could collaborate with Oaxacan artisans in order to experiment with materials and transform artifacts derived from the artisan community's cultural traditions and histories.

Initially, the team and I proposed that the workshops could be used for developing an object that would be produced by different artisan communities within the rural areas of the Mixtec Highlands in rural Oaxaca, known in México as a cradle of unique craftsmanship. The artisans' reputation is well earned through their uses of indigenous ancestral techniques in ceramics, woodwork, metalsmithing, leather, weaving and embroidering, and pre-Columbian uses of materials. During the initial process of ethnographic research, we facilitators held a round of exploratory visits to artisan communities, followed by initial narrative workshops in three specific communities: Santa María de Atzompa, Santa María Yavesía, and San Pablo Tijaltepec who work with wood, ceramics, and textiles. After these initial workshops, I discovered a multitude of logistical challenges that I discuss in detail during the next section.

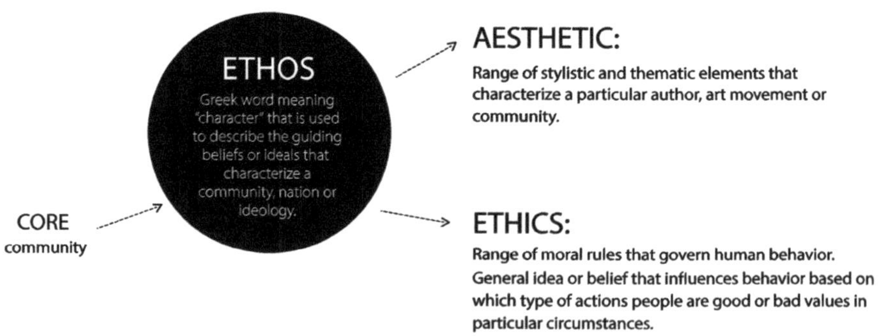

Figure 4.1 Infographic Community's core; Ethos, ethics and aesthetic designed by Rut Martín and Carmen Malvar. CADA Foundation®. Note: Definitions from Dictionary RAE (Real Academia de la Lengua Espanola) Royal Academy of the Spanish Language.

Our initial idea evolved through observation and learning more about the local reality, as well as evaluating the effect of collective action within these communities which offered the team and I a new and richer perspective. As I have detailed above, for communities that rely on non-written cultural transmission, form, color, and graphism become tools to transmit history and culture. These objects are tools for transmitting traits of individual and collective identity, and they serve purposes beyond practical applications. When applied to ritualistic or practical artifacts, these become cultural objects that transmit history, values, and symbolism. It is within this perspective that the material of oral tradition becomes relevant as it informs the transformation of created shape, imagery, and material manifestation as artifacts. We would later incorporate this into our narrative-based approach by "making with the hands" and established a set of strategies that were guided by the artisans' self-realized goals.

Our team's ethnographic observations of the artisan communities led us to ask ourselves new questions: Could a design identity be global without losing its local value and meaning? Can the intrinsic value of local tradition—that of specific craftsmanship as it is practiced within these communities—be maintained and also become global through a new design process? While the process was facilitated by design practitioners, our proposals during the workshops were based on the artisans' self-recognized needs, interests, and processes.

A core question during the formulation and implementation of this project was whether the aim of the workshops should be oriented around bringing the craft of indigenous communities to global attention and markets, and if so, if this could be done without sacrificing the integrity of the local culture. Although all of the communities actively sought the services and knowledge of global markets that we could provide, the question of potential risks remained a concern for the team and myself. There is a potential paradox in how an "a-capitalist culture," such as the isolated indigenous communities where the workshops were held, could access global markets.

Dialogue between facilitators and the artisans would be generated according to each community's expressed interests and needs, and we facilitators sought to avoid imposing any ideas or new directions. From the facilitators' perspectives and observations, and given the request communities' requests for increased market exposure, we agreed that an exchange of tools and skills through experimenting and developing artifacts together would mutually serve all participating parties. Not only would such a process provide the artisans with new tools and skills to achieve their goals to gain access to the global market, but their artifact creations would also retain and reflect the cultural and traditional traits of their communities.

The artisans' creations were conceived of and constructed from their own communities' histories and values. This project was ultimately concerned with creating a critical reflection through design spaces where transformation can occur during the processes of making together, and where indigenous artisans possess the agency to decide for themselves whether or not to transform their own processes. In this way, transformations to artisanal techniques and artifact construction are informed by the self-recognized needs of each artisan and their respective communities. In this sense, artisan indigenous communities can build resiliency against intrusive and aggressive global market and production mechanisms that scour the globe for labor forces.

During one project phase, which focused on prototyping artifacts at San Pablo Tijaltepec, the design facilitators shared several novel techniques for transforming the detailing and finishing processes of woven and embroidered figures. Some of the techniques introduced included reverse seams for figures and how to use a sewing machine for stronger, more even seams. Facilitators also introduced the use of patterns for improved size and shape consistency. We approached the introduction of new techniques with two goals in mind: maintaining our sensitivity toward craft without compromising its authenticity and unique handmade qualities, and centering the artisans' self-determination in the design and making processes. At all moments the artisans decided what to develop, what to market, and how to set prices for sale.

Facilitators offered, exchanged, and made contacts through their professional and personal networks, but the decision of what to do with these services and knowledge was under the total management and control of indigenous artisans and their communities. For example, one group of artisans organized a new collective, Colectivo Ñaa Ñanga of San Pablo Tijaltepec, and established their own online shop. Details of this organization can be found in the Chapter 5 field notes. I maintained my own political neutrality as an individual and designer in México by creating the CADA Foundation, which was intended to provide structure and legal support for current and future workshops and other initiatives. CADA maintains a coherent and transparent mission in contrast to the typical governmental efforts, and which allowed this project to serve its academic purposes while facilitating more trusting relationships between the team and each community. However, my decision to establish CADA and refrain from participating in local political or private discussions and conflicts, along with my efforts toward transparent communication, were not always positively received.

As outlined in Part 1 of this book, the theoretical foundation and goals of this project were developed through parameters for approximation and implementing a system of exchange with the goal of supporting local economies and identities. The first stage of the project involved facilitating, developing, testing, and communicating different "proximities through design," for which I conducted intensive field work and research in order to understand and contextualize historical, cultural, and social issues relevant to the communities I would later work with. Meta-research was documented and compiled extensively. Some video and audio recordings are mentioned throughout the document. I began my inquiries by considering the cultural backdrop, setting, and format for group exchange, which became my specific references to workshop activities and actions.

But, balancing human relations within cross-cultural approximation was difficult and intricate. Due to the fact that there is a strong tradition of Sistemas

Normativos Indígenas (once referred to as "Usos y Costumbres" during the 1980s-1990s),[1] the elders, who are mostly men, are in traditional positions of power within these communities. It was important for me—a güera[2]—to address any contact with the utmost care and by keeping a discrete and apolitical positioning. This is not to say that I acted as an unobtrusive researcher, but I rather maintained my position as impartial to the local and national political spectrums that rule over everyday community life throughout the course of this project. This also meant that it was important for me to establish transparent aims, goals, methods, and approaches to create a collaborative effort between other facilitators who were already in contact with artisan communities and the local authorities.

Sistemas Normativos Indígenas is a social and political tradition of indigenous customary law common throughout indigenous governance in Latin America. As I was familiar with a similar system of social relations and power structure through my own experiences in rural Galicia, I could relate to some aspects of Sistemas Normativos Indígenas in the Oaxacan context. In Oaxaca, 418 municipalities, out of 517, are ruled through Sistemas Normativos Indígenas, and it is granted a certain degree of autonomy within the Mexican Government. It involves aspects of politics, culture, approaches to work, economies, and sometimes communal ownership of resources. The communities with which we conducted our resources were governed through the Sistemas Normativos Indígenas system. Within these indigenous communities, leadership is seen as a form of community service organized hierarchically around a system of assigned roles and an Assembly. Municipalities are structured through

Figure 4.2 Visiting communities—questions of power, gender, and local authorities. Authorities communicating the arrival of facilitators at the first workshop. San Pablo Tijaltepec. CADA Foundation®, 2013.

a series of ascending positions called Escalafón based on community service. Representatives are nominated openly by the Assembly and without secret vote. This communal form of socio-political organization does not depend on the existence of, or participation in, political parties and is responsive to local cultural and ethnic rules. Roles, assemblies, and tequio—which translates to an indigenous notion of collective work, such as the construction and repair of road infrastructure, water wells, maintenance of municipal buildings, and organizing local festivities—give individuals within a community a sense of belonging and identity.

My first connection with the artisan communities was through a former student of mine who had contacts within the regional government and who liaised with local authorities. In turn, the local community authorities invited artisans to participate in our workshops through their own networks of communication. Community elders and Assembly members typically invited women to participate. The first meetings were held in communal spaces under the gaze of community leaders. At first, they also took part as curious, but mostly passive, observers. Working through this official network to invite artisans caused for the inadvertent miscommunication of our perceived objectives during the first workshops, and which made community organizers and participants believe that my research was being promoted and funded by the regional government. This ultimately generated false expectations for compensation among the artisans, and caused tension.

After this experience, we facilitators concluded that we must communicate the purposes of our workshops clearly and directly to the artisans from the outset. It was also important that the facilitators had previous contact and deeper understandings about the artisans and communities who opted to participate in the project because we could then build greater trust between all parties. We explained our expectations of the workshops and our goals to provide community support. We speculated with the artisans about the potential outcomes, and how practical matters could be addressed such as the provision of materials, potential monetary compensation for time or artifacts produced, food, transport, etc. Slowly, once the purpose and outcomes of the workshops were understood, community mistrust toward the team and I was reduced. This was also reflected in spatial arrangements: we moved from wider communal rooms into private spaces, and we were invited into the homes of women who wanted to take part in our workshops out of their own motivation or curiosity.

Under Sistemas Normativos Indígenas, the authorities still had to approve of our activities and give the women "permission" in order to engage with the workshops. Once we began the final workshops at San Pablo Tijaltepec, however, the team and I recognized a substantial change in men's attitudes and activities, and they even began to take charge of domestic activities that usually lay in women's hands: i.e., preparing food, housekeeping, taking care of children, etc. Remarkably, this shift occurred so that the women artisans could participate in the Second Mesoamerican Textile Congress (TEXTIM II) and Expo-Sale at Oaxaca de Juárez held by the Oaxacan Textile Museum.

Overall, the team and I recognized a major shift in the attitudes of both men and women in regards to women earning money without involving men in their environments and activities.

After the initial workshops, the team and I held meetings before and after each ensuing workshop where we would discuss our reflections. During the workshops themselves, we conscientiously documented events, recorded participatory activities and reactions, annotated the type, occupation, and use of spaces, and recorded participation by groups of people, such as the involvement of elders, community leaders, women, and children. We facilitators explained the purpose of this documentation process to the artisans, and they granted us their consent before we began. After each workshop we collectively reflected on the recordings of outcomes and the artisans' perceived emotional responses to activities: interest, curiosity, joy, trust, suspicion, and expectations. We also recorded the work patterns, behaviors, and social mores demonstrated materially and through language. In the group dynamics, particularly, I took note of the expressions, beliefs, and concepts communicated within the collective act. This process opened a well of information to us facilitators that we could use to be critical toward our work and make necessary transformations. During team meetings, we reflected upon the recorded design processes with regard to our observations of power, hegemony, and trust in the communities (Creswell 2013, pp. 94–95). The team and I also sourced materials, which I personally funded, and identify contacts who could contribute to a particular area of expertise for the community and the artisan collectives.

Over the course of the workshops, we worked together to generate a more equitable dialogue between local communities and global makers to result in the collective production of an artifact serving as an ambassador of this dialogue through mutual exchange. In my experience as a design practitioner, ideation, abstraction, and conceptualization happen before the act of making through prototyping or study models. Artisans, on the other hand, create artifacts foremost through manipulating material in reference to a deep knowledge of materiality. Artisans possess this intrinsic knowledge that is learned during childhood, and they develop changes through numerous iterations resulting in slow transformative patterns. Through a process of reflection and collaborative design practices, this project created a space for exchange between artisans and design practitioners while also directly supporting the goals of artisan communities to gain exposure to new markets.

My approach to fieldwork addressed some of the open issues at the heart of design theory and practice. The practice itself involved a form of reflective communication for social design, which is based on my understanding of design after years of practice as both a student and as a professional. I characterized it through its potential for transformation at the social and economic levels and its ability to strengthen cooperation and democratic values. Within the context of this project, it was our task as facilitators to trigger collaboration between design practitioners and artisans in creative environments, not to solve any single design "problem." With this approach, our fieldwork identified communities where future workshops could be held and gathered a team of like-minded local design practitioners who would become a team of facilitators for necessary community service.

Building a Team of Reflexive Design Facilitators

Gaining the respect and earning the trust of the communities we engaged with was a slow process that was developed over a long period of time through relationships between individuals and over the four years during which workshops were held. This slow process was reflected in the way we related to each other—greeting, addressing each other, the use and perception of personal space, the manifestation of personal trust and appreciation. I was aware that, in the past, other designers had approached indigenous artisan communities to conduct isolated projects without continuity. There were very few instances of long-term commitment to projects, as the collaborative project in Chiapas initiated by French textile designers who worked ad created together with the community, the organization called El Camino de los Altos.

Through a social design process embedded within practices of correspondence, the team and artisans developed social relationships across time and by means of open dialogue, active listening, analysis, understanding, and transforming workshop proposals. Through using methods rooted in *proximities of design*,[3] we could develop a collective dialogue and enact democratic participation in decision-making during workshop preparation with facilitators, adapt and evolve the processes during the workshops themselves, and keep open a process of mutual transformation for both facilitators and artisans. We facilitators strove to develop a common objective with the artisans, to refrain from imposing predetermined materials and outcomes, and to listen and respond to artisans' needs and interests. All the while, we worked to document and reflect upon outcomes to determine steps for the next phases of workshops.

Our proximities of design methodology proved that local sensitivities had to be navigated with care. My approach to social design constituted one of the pillars of my work and a main topic within all the facilitators' workshops, with the aim of maintaining a position of political neutrality toward the national and local political or economic agendas. This position also required me to establish and communicate effective and transparent goals, interests, and objectives during the workshops. The team and I also expressed our desire to develop continuity in future workshops within these, and other, artisan communities.

This project was predominantly conducted with two different participating indigenous artisan communities in the Mexican state of Oaxaca. Myself and six other professional designers—acting as facilitators—developed the workshops, which took place in phases two or three times a year between 2014 and 2017. Our research approach in Oaxaca was twofold: on the one hand, we maintained a theoretical approach to integrative approximation, and on the other hand, we maintained a practical approach to fieldwork as a method for understanding and contextualizing situations through developing personal relationships between designers and indigenous communities with empathy, reflective observation and action. This project would not have offered the same results if my initial connection with the communities had not been made through contacts deriving from personal, professional, and academic networks. These networks eventually led to the gathering of a group of insightful and talented local design practitioners with experience and deep relationships at the local community level.

My journey to build the design facilitator team began by contacting local authorities through an Oaxacan designer and alumni Fátima Díaz, who had connections with a regional government official. I also connected with a fellow academic in New York who advised contacting Maddalena Forcella, a textile designer working in Oaxaca. Forcella, an Italian textile designer, has lived in México since 1987. She was involved in the Camino de Los Altos collaborative project, and her experiences with the collective provided this project with valuable knowledge while we developed our workshops and community

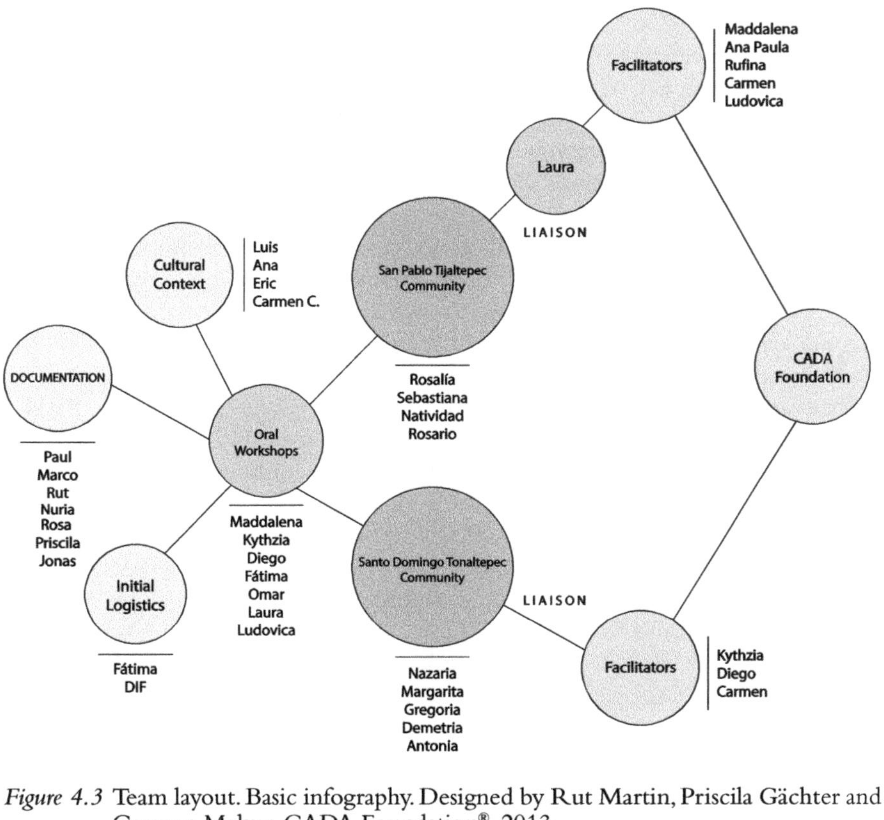

Figure 4.3 Team layout. Basic infography. Designed by Rut Martin, Priscila Gächter and Carmen Malvar. CADA Foundation®, 2013.

relationships. She also had previous and extended experience in archaeological restoration in Italy, having trained as a textile designer in Mexico City, and worked with textiles in San Cristóbal de Las Casas, Chiapas. There, she co-founded K'un, a pioneering workshop for Mayan weavers that introduced new designs and contemporary sensibilities to the artisans' production of textiles with traditional techniques using natural dyes.

Maddalena's perspective was quite enriching for the project. She had exchanged experiences and knowledge with weavers from different indigenous groups, taught classes and qualifying courses in Chiapas and Oaxaca for both government institutions and civic groups, as well as for associations in México and Guatemala like Fondo Nacional para el desarrollo de la Artesanía (FONART) and Programa Nacional de Culturas Populares e Indígenas, Conabio. Her previous work in the design and the production of artisan objects with national and foreign companies—including Artefacto, Arroz con Leche S.A. in México and Iris Von Arnim in Hamburg, Germany—and her directive role as Head of the Craftsmanship Development Department at Museo Na Bolom, in San Cristóbal de Las Casas, Chiapas, provided Maddalena deep insight into the dynamics of creative and productive processes for the Oaxacan context in particular. *Meta-research: See video of Maddalena Forcella: https://vimeo.com/489889314/1fe3c52389.

Her experiences revealed close ties to weavers that were developed through time, trust, and respect, which she had gained as an initial collaborative member of the Mayan weavers and designers association El Camino de los Altos.[4] Upon meeting with Maddalena Forcella, she referred me to Anthropologist Luis Morales[5] at San Cristóbal de Las Casas. Long conversations between the three of us followed topics of Mexican history, cultural development, indigenism, anthropology, and the role of craftsmanship in oral tradition through intergenerational, family-led, informal training, and regulated training. We also spoke about the role of CECATI—Centros de Capacitacion para el Trabajo Industrial Government training centers for industrial work—in training generations of craftspeople for "industry-oriented jobs."

Through Luis Morales, I was also introduced to artist Ana Pellicer, a well-respected sculptor and co-founder/former Director of CECATI 166 Best Maugard in Santa Clara del Cobre. She and her late husband James Metcalf—a well-known American artist and master metalsmith—established the school. Their school had earned a reputation as one of the best training centers in the country that focused on metal craftsmanship.

In due time, Maddalena Forcella also introduced me to Ana Paula Fuentes, a textile designer and former director at the Museo Textil de Oaxaca. Ana Paula Fuentes, born in México Distrito Federal, had established a collaborative design practice in Oaxaca. Trained as textile designer and an expert in knitwear and fashion design, she had vast experience working for Spanish and Mexican fashion brands with both a local and international perspective. Co-founder of the experimental 11011 Studio in Oaxaca City, Fuentes collaborated with more than 20 contemporary artists organizing exhibitions, performances, installations, and concerts. As founder of Artesana, she had created a new craft product market network spanning across México, and had finished a six-year tenure as Director at the Museo Textil de Oaxaca (Textile Museum of Oaxaca) in 2012, where she maintained strong ties. Fuentes had developed, promoted, and marketed crafts for La Flor de Xochistlahuaca, a cooperative of 37 indigenous Amuzgo weavers from Xochistlahuaca, Guerrero and collaborated with Eric Mindling in formulating and conducting study travel tours for international weavers, scholars, and the general public interested in visiting Oaxaca's rich cultural and vibrant artisan communities. *Meta-Research: See video of Ana Paula Fuentes: https://vimeo.com/489888880/a05b5371ab).

Fuentes then introduced me to Eric Mindling, an American photographer and author of a book about Oaxacan pottery. Mindling is also a private ceramic collector and an expert in Mexican traditional pottery. A founding member of Innovando la Tradición, a creative and collective platform where craftsmen, designers, and artists shared skills, knowledge, and stories, he had documented the pottery practices in 65 different communities in Oaxaca. This work is reflected in his reference book, *Barro y fuego; El arte de la alfarería en Oaxaca*, which was the product of a two-year long research process and offered a panoramic vision of Oaxaca's pottery past and present, along with insights into its possible future. The research team also continued to grow as Eric Mindling brought along two other designers and ceramists, Kythzia Barrera and Diego Mier y Terán, who were members of the Colectivo 1050 artisan cooperative.

Kythzia Barrera, an industrial designer with a strong and innovative approach to social and sustainable design, was co-founder of both Innovando la Tradición and ceramists' Colectivo 1050, where she had succeeded in supporting the development of pottery communities around rural Oaxaca. Her personal approach bridged art, craft, and design to foster social and human changes with a focus on sustainability. She and Diego Mier y Terán shared this approach, and their methods privileged an honest approach and open discourse in their collaborative work with artisan communities. This contrasts the more conventional designer-centered methodologies. *Meta-research: See video of Kythzia Barrera: https://vimeo.com/489879902/09bbf58a88.

A graphic designer and university instructor interested in social and ethical dimensions of design and its power as an agent of change, Diego Mier y Terán ran the Utopia workshop at the prestigious Universidad Iberoamericana on the topic of Philosophy of Design. He believes in the power of design to transform Mexican Society. Another co-founder and director of Innovando la Tradición, and member of Colectivo 1050, Mier y Terán collaborated at Estudio Frutas y Verduras in the fields of graphic and editorial design. Diego brought a sharp mind with a deep and critical train of thought, and volunteered valuable perspectives during facilitator workshops. He often contributed reasoned opinions about collaborative structures, artisan workshops, co-design processes and cultural approximation. *Meta-research: See video of Diego Mier y Terán: https://vimeo.com/489893081/2e2d02ad46.

Diego and Kythzia would eventually introduce me to Omar Fernández, an architect with experience in woodworking.[6] He joined the team for a while, as he had a standing collaboration with a group of wood artisans at Santa María Yavesía. Eventually his professional agenda would prevent him from participating more fully. Although he was forced to reduced his participation, Omar remained an open-ended collaborator with common interests to this day.

Another key collaborator at San Pablo Tijaltepec was Laura Quiroz. Trained as a design engineer, she became the main liaison between ourselves and the community of textile artisans. Her previous research at Centro de Diseño de Oaxaca for the Universidad Tecnológica de la Mixteca positioned her as a documentalist, having previously produced documentary videos about crafts in Oaxaca as a tool for inter-cultural dialogue from a community perspective. She also had directed the documentary audio-video *Naa kiku isaa y kiku sama* (Spanish translation: Mujeres tejiendo nuestra cultura, English translation: Women weaving our culture) which showcased both the production process and design elements in traditional textile works of San Pablo de Tijaltepec used by the female artisan community. Laura Quiroz would also become a powerful voice in defense of the interests and copyrights of artisan communities and was involved in requesting Patrimonial Protection on the San Pablo Tijaltepec embroidery, symbols, and techniques. *Meta-Research: See video of Laura Quiroz: https://vimeo.com/489889708/e8ff439d66 *Meta-research: https://vimeo.com/89625876.

As practitioners, we all had a common interest in design as a practical action that was also grounded in our years of experience with collective projects within communities of people from different backgrounds. In our first meetings as design practitioners, we shared our experiences from previous projects. We were all invested in reflecting on the process through examining the material practices of "doing design."

Other local collaborators had more discrete yet equally important roles. They liaised with community leaders, assisted as photographers, recorded events, acted as translators, and facilitated workshops with children. Others were impromptu chauffeurs, interpreters of local customs and languages, solvers of practical matters, or acted as mediators when circumstances got tense. These collaborators were, among others: Eric Chávez,[7] a fourth-generation weaver, and collaborator at the Oaxacan Textile Museum; Ludovica Morales, a young graphic artist; Paul Mijangos, a local photographer and communicator[8]; Hugh Atkinson, a colonial and post-colonial historical researcher; Emmett Werbel, a political speech writer and editor; and photographer Marco Valentino, who also did some video coverage and documented some travel and workshops. I also acknowledge the valuable participation of anthropologist María del Carmen Castillo and architect Rut Martin, fellow colleague, professor, and a keen researcher based in Barcelona. She collaborated by offering her insights and managed key documental organization, infographics, recordings, and took part as facilitator at onsite workshops and meetings with communities during Spring 2014 and Summer 2017. After our conclusion of the workshops, I also recognize Nuria Widman, designer and researcher, and Jonas Adams for assisting

Figure 4.4 A team meeting at the café center in Oaxaca. CADA Foundation®, 2015.

me with editing the final versions of this project and with the publishing process together with Priscila Gächter and Rosa Chapela getting edited footage and final selection of illustrations for this publication.

All aforementioned collaborators played different roles in different stages in the development of this research project, and offered me a unique framework perspective, as well as an understanding of previous or current collaborative structures between researchers, artisans, and communities. Their insights provided me with a better understanding on how to navigate and establish real co-creative communities through the organization of a functional facilitator team and our development of collaborative workshops. As I worked with this group of professional design facilitators, we shared collaborative experiences as participants in the workshops. Our experiences shared with the artisans helped us all to build a relevant and solid understanding of the local social and cultural panorama.

Selecting Communities Through Team Connections and Discussions

Once the designer facilitator team had been established and our first meetings conducted during the month of October 2013, we jointly agreed to propose an approach through oral narratives to the artisan communities as a way to create dialogue through their intangible artifact as a vital part of local culture, history, and cosmological vision. Narrative, in this context, has been considered immaterial cultural heritage by the United Nations Educational, Scientific and Cultural Organization (UNESCO).[9] UNESCO states:

> Intangible and cultural Heritage, transmitted from generation to generation, is constantly recreated by communities and groups, and provides them with a sense of identity and continuity, thus promoting respect for cultural diversity and human creativity. (UNESCO Convention for the Safeguarding of the Intangible Cultural Heritage.

Us facilitators met and shared our first approximations during the oral narratives as a means to approach the community through their intangible artifacts. We shared our ideas with those artisan communities who had requested support for new market exposure. Design workshops were then conducted in the indigenous communities of the Oaxacan state in México, whose creative and artisanal processes risked disappearing. We led a series of exploratory visits at five communities, out of which we conducted initial workshops at three: Santa María de Atzompa, Santa María Yavesía, and San Pablo Tijaltepec. During facilitator meetings after our explorations, we outlined a process of design workshops to empower the social and political structures of the villages. We facilitators and artisans held a second round of workshops in San Pablo Tijaltepec and also incorporated the community of Santo Domingo Tonaltepec due to their request for market access and marginalized social situation. These workshops consisted of oral narratives expressed by attendees through either drawings or small modeled figurines.

Lack of accessibility served as an important selection criterion for the initial exploratory visits. Though the challenges and difficulties were many, we agreed to focus on those difficulties as temporal experiences that could inform the design process. We recognized issues defined by isolated geographic landscapes, logistic frontiers, and obvious language barriers which led us to consider questions of power, gender and local authority.

Over a three-year period, we realized that it was fundamental to bringing into view the politics of design, including the systematic placement of politics beyond the limits of design's frame. As Margolin (2002) observes, "if designers are going to realize the full potential of design thought, then they should also learn to analyze how the situations that frame design practice are themselves constructed."

Another crucial criterion for selecting initial communities was their desire on initiating larger market exposure, the degree of product exposure to outside influences, market access, the threat of unique technique disappearance. For pottery artisans, facilitators initially looked for communities who created smaller pieces. Eric Mindling and Colectivo 1050 suggested Santo Domingo Tonaltepec for the artisan's pre-Hispanic technique using resin (tree sap) and their marginalized political and social situations. Maddalena and Ana Paula added that the town of San Pablo Tijaltepec should also be considered due to

Figure 4.5 A permanent inaccessible road section. Santo Domingo Tonaltepec. CADA Foundation®, 2015.

their unique handcrafted embroidery technique, called pepenado, where facilitator Laura also had ties. Eventually, we facilitators narrowed the project's scope to include two Mixtec communities: Santo Domingo Tonaltepec and San Pablo Tijaltepec, both located in the Oaxaca Valley.

In these communities, we facilitators and artisans focused on developing prototypes based on cultural artifacts as both the catalyst and mediator of the project. These two communities had entirely different social realities and ways of expressing them. However, on the geographical and cultural level, they represent a common realm in which the artifact plays an important role in the social fabric of the community, and is an essential part of social life and witness to changing contexts and social transformation.

Santo Domingo Tonaltepec is one of the few communities where artisans still produce the same way as in the pre-Hispanic period, and thus, Hernández (2011) believes it to be an excellent example of cultural continuity. Geographically isolated, Santo Domingo Tonaltepec retains its idiosyncratic ceramic practices. In three of the community's villages, including Río Blanco Santo Domingo Tonaltepec, the pigments come from the bark of certain trees which are rich in tannins. They use oak bark to produce a pigment that is then poured drop by drop over the clay pot or jar while it is still hot from the oven. The liquid crackles as water evaporates on the hot clay surface and leaves only the red russet color of the tannins (Mindling, 2011). The production of this pottery materializes several dimensions of the artifact's cultural substance: the organization of the production, distribution and, the use of the artifacts themselves (Arellanes Meixueiro and Simon Reyes, 2011; Hernández Sánchez, 2013).

In San Pablo Tijaltepec, artisans produce a variety of fabrics and traditional designs through embroidery and pepenado, or a method of pick-up weave and gathered pleats. Historically, women produce fabrics for either domestic or ceremonial use; and knowledge of this craft is passed down from mothers to daughters (Turok, 1988; Ortiz Escamilla, 2012). San Pablo Tijaltepec textiles display elaborate graphic and figurative designs with great mastery of scale. Without any written directions, these textile goods are produced with a high degree of sophistication. As artifacts of cloth and thread, they hold special significance not only as social objects, but also for their deep personal nature. As Peter Stallybrass observes, clothing "has an indisputably intimate relationship to persons—not just their appearance and social identities, but even their gestures and smell" (Stallybrass, 1996, p. 183). Therefore, it was crucial to pursue approximation with reverence for the individual character of these artifacts.

Although different in social and cultural contexts, there are common threads between both communities: residents of Santo Domingo Tonaltepec and San Pablo Tijaltepec live with precarious resource access, poor communications, and a high level of political and economic marginalization (Ortiz Escamilla, 2012). Upon seeing the artisans' craftsmanship and the situation of local artisans within both communities, we facilitators considered it critical to acknowledge textile and ceramic as materials which, once transformed into artifacts, would go on to perform deeply personal and critical functions.

The traditional artifacts of these communities are still crafted using pre-Hispanic methods, although many of their shapes and designs have been transformed significantly over time (Hernández Sánchez, 2013; De Avila, 2013). While some artifacts have kept their original dimensions and forms, others have been changed. What remains important to the artifact is their production process and their uses (Balaram, 1995). The artifact production has always played an important role in the rural life of these communities and their identities (Pellicer, 1995). This is because the artifacts are representative of communal history and culture through the artisanal processes of making and conveying knowledge. Artisans may transform or changed artifact designs over time, but the traditional making process remains as a crucial means to convey and preserve knowledge. Thus, considering the complex role these artifacts play as historical witnesses, the artifacts of these communities can really be considered cultural "time capsules" (Hernández Sánchez, 2013).

Process as Method: Learning and Transforming Together

The development of our project methodology was based on experimenting and transforming materials together, collective reflection on artifact production and design, as well as learning about oral tradition and its role in cultural transmission. This involved learning about local ethics and aesthetics through an iterative, organically evolving process that we facilitators carried out as reflective practitioners.

Throughout the entire course of the project, we facilitators remained critical of our own design processes and supported artisan goals. This included initiating larger market exposure for both artisans and their wares, as long as that was their desire. We held additional workshops with artisans to offer our support and provide materials to finish samples, refine finishes and prepare objects to be marketable at Museum Shops and textile events such as the TEXTIM II and Expo-Sale at Oaxaca de Juárez. Other practitioners provided insight into marketing strategies to approach a wider, global audience per the artisans' requests for such support.

As I have mentioned, the initial phase of this project involved extensive fieldwork and exploration before planning each workshop. We facilitators focused on ethnographic research, both in theory and through our field work, as we met key figures in the development of craftsmanship and training, anthropologists, artists, and artisan communities. I interviewed ethnographers, artists, anthropologists, and made visits initial to selected communities. First, these served as exploratory visits to follow up on recommendations made by those who had knowledge and connections within the communities, and later these visits led to our development of one-day workshops in selected communities that expressed their desire for strategies to attain market exposure. The second phase was dedicated to knowledge exchange and design experimentation, broadly understood as collaborative and a transformative design process. Our main goal as design practitioners was to develop scenarios for long-term engagement with communities, and establish a methodology of approximation and processes through 3D exploratory artifacts. The third and final phase of research focused on developing prototypes for artisanal production through iterative workshops. During this phase, our group of design practitioners worked alongside artisans in order to continue developing the most effective methodology while generating a series of samples. Below, I discuss some of the background information pertaining to each phase that is necessary to set the stage for the case study field notes provided in Chapter 5.

Phase I: Initial Ethnographic Research and Oral Workshops

We began the initial Ethnographic research by focusing on the social, political, and cultural background which shaped circumstances in the participating Oaxacan communities. Several ethnographical issues were researched through an extensive analysis of literature: the ethnography of spoken language—of particular interest in Oaxacan culture due to its important oral tradition—as well as the relationship between aesthetics and ethics in Mixtec context.

Oral tradition is the vehicle for cultural transmission, becoming itself an intangible artifact, and how artifacts become cultural objects as they transmit cultural history, values, and symbolism. It is under this perspective, that speaking—as the "material" of oral tradition—becomes relevant. Due to the importance of oral tradition, and to ground an important piece of the fieldwork, I set out to interview a poet and writer focused on studying and recording cultural manifestations in Chiapas. One of the persons whom I first met and interviewed was the artist and activist Ámbar Past who also founded Taller de Leñateros[10] (The Woodlanders' Workshop). Past was born in Durham, North Carolina, United States, in 1949, and eventually made her home in the Highlands of Chiapas' rural hamlets where she learned to speak Tzotzil Mayan. She is culturally active, and participated in the formation of The Mayan Writers Centre in San Cristóbal de las Casas—a cultural society established in 1975 aimed at strengthening, highlighting, and spreading values of the Amerindian popular culture including songs, spells, and legends. She has also helped organize and run other cultural and artisan groups, such as Sna Jolobil—a weaving cooperative for Mayan artisans—Sna Jtz'ibajom—the Mayan writers collective—and presided over the Libros Prehispánicos A.C. literature association.

Past has also served an important role as a collaborator of Sociedad Indígena de Chiapas,[11] a cultural institution aimed at preserving control over its own cultures and achieve a creative state of peace and order through security and food sovereignty, justice, and dignity. The organization and its aims were established in the 169 OIT agreement, and under the terms of events around self-determination in the region. Its mission is to:

> Promote and support activities that strengthen the integral development and the quality of life that Chiapas Indigenous people deserve, as well as join in the scientific and humanistic projects and programs that work on the humanity's settlements by means of a deeper relationship with Mother Nature that ensures and improves life on our planet.
>
> (Sociedad Indígena de Chiapas)

Past's activities involve, within the local context, taking a political stance as well as working as an activist in civil associations. Her first writings were published in Tzotzil in the collective book *Slo'il jchiltaktik* (Autobiographies of Tzotzil Women) (1978), and *Bon* (1980), a bilingual Tzotzil-Spanish manual for Mayans on natural dyes. Past has also published several books of poetry in Spanish. For the past 30 years Past has collected, recorded, and translated Tzotzil ritual poetry and published in bilingual anthologies by Taller de Leñateros: *Conjuros y ebriedades* (1998), *Incantations by Mayan Women*, a music CD-book *Disco de los Conjuros*, and, *Bolom Chon* (in press). References to these works and publications, among other supporting materials, were used as models to develop the oral workshop methodology in this research project. Other interviews, along with notes and videos documenting my interviews and meetings with social designers working with artisan communities in San Cristóbal de Las Casas and Oaxaca, offer valuable perspective into my approach and methodology development. I discuss these meetings in detail during Chapter 5. ★Meta-research: See Vimeo of Ana Pellicer and Luis Morales: https://vimeo.com/489872162/c084d5f4e8.

Based on our initial research and facilitators' previous experience with communities based on oral tradition as cultural transmission, we determined that our novel methodology could be implemented in a series of Oral Workshops that would analyze, interpret, and evolve cultural artifacts based on oral popular narratives. According to Ornelas and Gregory (2011), "A Cultural Artifact is defined as an artifact—tangible or not—that has embedded meaning for a socio-cultural group and therefore has potential to elicit emotions in individuals." Thus, the oral workshops were useful as a way of relating community and artifact through oral tradition and using it as an expression-generating element that the workshop's participants could materialize into drawings and graphic expressions. Drawing is a means of communication equivalent to writing: it is a language used to tell a story. The workshop drawings showed reality and the collective imagery of the members of the community, and served as the articulating medium between facilitator and artisan to interpret and produce the tangible artifact that we were striving to reach. In this way, the physical artifact would be the result of a transformation of the community's expression, or its ethos. Tactically, as design facilitators, we agreed to conduct workshops with several generations of mainly women and children who expressed their oral narrative through drawings. Then, after reflecting on each workshop, facilitators met and made selections from a series of drawings that served to develop common characters and stories between participants. This was the starting point for the project's next step, in which the tangible characters/elements were devised using each area's specific crafts to produce the different pieces.

We prioritized eliminating the labels that drew distinctions between popular art and design, and instead we focused on "forms of making" as a way for communities to make their cultural and social value systems more tangible in physical form. We thus focused on the semantic language of artifacts within both indigenous communities, while also reflecting on our own perceptions of the chosen images. By undergoing this process for selection, we discovered personal tastes linked to participants' own cultural references.

We facilitators then analyzed the drawings created by the communities. The materials selected afterward were, during the design phase, the articulating bond in maintaining open dialogue with the artisans. It is important to stress that, in this context, the artisan belonged in the same community where each of the workshops took place, and therefore each artisan shared the same "ethos" bond with their respective community.

Phase II: Design Research and Transforming Methodology During the Making and Unmaking

Between our completion of the initial oral workshops in Phase I and our beginning of Phase II, the team and I met to reflect and discuss our next steps. While considering ways that we, as facilitators, could be supportive of the artisans within an applied design methodology, we asked the questions: How can social design generate a positive effect within an established value system? How can we think in terms of a culture of transience governed by local stories rather than the subordination of local culture to global design? As Tunstall (2013) explains, such a system must be accepted as dynamic and that all parties involved should assume within which there will be a value exchange among different cultures and systems. Furthermore, she emphasizes the importance of eliminating any circumstances that generate a situation of disadvantage during exchange itself.

We facilitators were aware that we were dealing with human beings who were not going to leave their emotions, duties, and memories behind when they came time to collaborate. Ranjan (2011) observes how social design, when put into practice and subjected to democratic ideals, can become a powerful social and political activity that gives voice to local people. We could not pretend to address this process in an abstract or "global" way. We had to address the problems and challenges at the local level where the exchange happens, and with the context informing our collective practice.

Consequently, we organized the work around a common project. This created an informal structure to transcend public–private distinctions, relying on collaboration rather than competition, and a new form of communication. We raised questions and issues from the workshop and used these reflections to inform our logistical discussion of the next meeting place, time, and day. The project invited practitioners to use their understandings of design to allow us all to enter in a form of community with the artisans who also possessed a different form of design knowledge. Throughout the project, we believed that deeply committed human relationships should guide our behavior during the workshops, for they generate a diversity of actions and solutions which are invariably strengthened through the exchanges and dialogues of the relationships. Our cohabitation was strong in the sense that empathy and participation allowed us to coordinate decision-making with more effective actions and promote accountability. When people from a variety of backgrounds form local relationships around shared tasks, trust can be bolstered and the social fabric strengthened (PROADA, 2002; Lazcano, 2005).

In his paper, *Product Symbolism of Gandhi and Its Connection with Indian Mythology*, which describes the spiritual and material aspects of design within the Indian context, Balaram (1995) discusses the power of the artifact's narrative over its function in specific regard to everyday objects. Rather than solely deconstructing form, Balaram insists that the narrative must be understood through the artifact's characteristics of manufacture, morphology, infrastructure, and decoration. He writes:

> All of us could be said to be at home in different territories [. . .] according to our experience of familiarity with them. Those using the same [. . .] share a common territory. The territorial imperative of semantics suggests that people who live in the same territory more readily accept each other's

> messages than those belonging to different territories. This imperative implies that people by themselves do not just talk to each other. The symbolic qualities of the material things they associate with always speak with them.
>
> (Balaram, 1995, p. 135)

Therefore, social design is justified as critical to understanding communication and inter-relation within communities. In his analysis, Balaram (1995) notes that communication neither emerges from isolated systems nor is it received through one isolated channel; he argues, rather, for the importance of the "multidimensional message" defined as a message that is delivered through multiple means, especially aesthetically, and is comprehensive in its totality. Artifacts of everyday use that carry the socio-cultural stories which are, perhaps unknowingly, passed down through generations can therefore be understood as objects carrying a "multidimensional message."

During Phase II of the project, we facilitators recognized "proximities of design" as moments and situations during the workshops where interactions between people, places, and things occurred within the frame of space and design. The team and I reflected on these moments of interaction during "the making" process as central to the artisan's expression of their own worldviews and goals during the designing and construction stages of their artifacts. We relied on approximations of the collaborative making process that also aimed to connect local context to the artisans' creation through reflections of their own techniques and interactions with space and materials. We started this journey by reflecting on the very theoretical framework for understanding activities, identities, artifacts, and communities of design practices. Could the practice of design help us better understand local identity, and how could we create an environment in which correspondence could be generated? We were aware of the importance of engaging with all the senses and in all dimensions with the environment we were going to share. Adams (2014) widens the scope of the current debate about participation through correspondence, defining this correspondence as inclusive of the spiritual and cultural values that constitute the community's worldview. According to Recchia (2011), the only way to immerse oneself in the geography of values is nothing other than to create a dialogue.

We facilitators asked the group of artisans—who showed both interest and uncertainty in equal measures—how they believed we could best work together. Many of these discussions within the project revolved around our invitation to all artisans who had expressed their interests in gaining market exposure. Together, we discussed further possibilities and, in this way, jointly decided on how we could deliver the results of our design process back to the communities and markets. The few women who initially participated showed a playful curiosity and a strong ability to improvise. We then moved from an open public space to more private working environments. These meetings and workshops are detailed as case studies in Chapter 5.

During the workshops and our post-reflections, we facilitators considered spatial limits and geometry, form, and repetitions of the same stories. This exercise provided an important view of how stories and collective memory differ: some did not focus on the stories themselves but rather followed their own imagination according to visual memories, while others experimented with the medium itself. We tracked and measured what we could about the traditions of each community's craftsmanship and how these traditions were passed from the old to the young. I have previously mentioned Assmann (2011), who emphasizes how the making process relies greatly on each community's own mechanism for transmitting cultural knowledge between and across generations. These mechanisms are dependent on cultural artifacts and traditions as ways to remember. Collective memories contained in the artifacts, or represented by their designs, also serve as a template to organize present and future experiences.

By the end of the second day of the initial Phase II, the participants had selected their local stories and transferred them to a new dimension of narrative that would inform the design through a 3D experimentation with materials. As dimensionality and scaling connect to participatory design traditions, diverse representations (texture, geometry, surface, material etc.) were necessary for conveying different aspects of design. Facilitators worked with artisans to make important design decisions about transitions between representations and scales. The new physical artifacts were then coded in a language that could not be separated from the medium itself, and communicated complementary aspects of the design process. We identified strong narrative elements in the ways in which design representations were combined and how the artifact was experienced from different scales and angles to induce various scale models sometimes using different materials and techniques (Telier, 2011).

Facilitators and artisans started the third day of workshops by discussing the project in all its aspects as a group. Several participants said they interpreted the activities in different ways than we had anticipated, and were unsure about various aspects. Artisans shared their insecurity over experimenting with new forms and materials, and we facilitators suggested that they refer to those that they already knew. The complexity of the embroidery patterns and ceramic processes was also a significant challenge during workshop discussions. This became particularly clear during facilitator reflections where we reviewed previous artifact designs. At this point, the project had gone on for months. Then, the artisans and designers gathered so that the artisans could select some objects and references different from those worshipped in the iconography of the artisan's own community. During the early stages of the project, I had built a collection of images to serve as alternative sources of inspiration during the workshops. The artisans and facilitators gathered to select some objects images as references based on their own tastes and preferences, and different from those that were worshipped within the wider community's iconography. The images were based on tridimensional toys representing artifacts chosen from different

Figure 4.6 Artisans of San Pablo Tijaltepec observing chosen images, printed in black and white from universal artifact selection gathered by myself. San Pablo Tijaltepec Design Workshop. CADA Foundation®, March 2015.

materials, scales, and cultures. We displayed the images on a wall during the third workshop and each artisan selected images that inspired them the most.

Based on our collective interactions with images of objects from a variety of different cultures, we facilitators asked ourselves: How and when can we consider the appropriation of foreign elements and processes valid? Bonfil, Novelo and Acha largely agree that appropriation is valid insofar as it answers an internal need to adapt and is introduced according to the agreed-upon needs of a group with communal cultural affinities. They all point to how communities themselves select which elements to maintain, incorporate or substitute over time. As shocking as this acculturation process may seem, it nevertheless resolves itself naturally to produce well-integrated forms and results. However, the Phase II workshops went one step further to generate a cultural exchange and a potential transformation of the material. Knowledge was combined, and elements originating from various heritages and contexts were incorporated together during the traditional making process.

Many understand crafts as the pioneer form of what can be considered the profession of design today. Historically, the practices of these crafts were relegated to the background under the pressures evolving industrial processes.

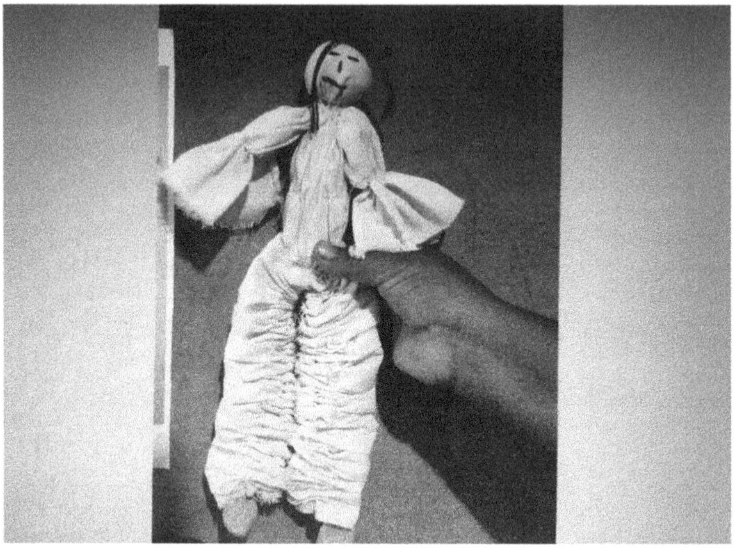

Figure 4.7 San Pablo Tijaltepec, artisan Rosalía Bautista. Cloth dolls made during workshops. Based on external references using "pick-up, pleated and ruffled" technique. San Pablo Tijaltepec Design Workshop. CADA Foundation®, March 2015.

Hence, creative acts and the associated skills of el hacer—"the making"—have been separated, with the common conception of the artisan becoming more associated with virtuosity than with the activities of design (Makovicky, 2011). From the beginning, then, we facilitators wanted to materialize stages, jointly speculate on the possibilities, and consciously avoid judging the artisans' work from a western value perspective.

An artifact has to be positioned in a particular setting in order to make sense, and for those existing beyond the respective value system to realize the intention and vision of the artisan. How then does the process of making unfold, then, as the artisan engages with objects totally different from those that worship the iconography within the artisan's community? Although the objects and figures do not change, their meanings experience a shift when they are transferred into a different cultural context. Kuchler (2011) studied the transformative effect of an artifact's production being moved from one material to another, and as it alters work relationships and everyday life. There are elements of iconography and worship in the artisans' communities whose meanings have not changed since the prehistoric times. That said, some objects are transformed when they enter or are taken into a different cultural context. Within the cultural context that artisans are familiar with, then, aspects of foreign designs can also be changed and appropriated during the making process.

Through research and navigating social design, the cultural and emotional values of the communities involved should be maintained. This becomes a challenge when new and exploratory artifacts begin to integrate the necessary aspects to appeal to global commerce. In our explorations, the artifact served as the articulatory element of the process. I also observed how organizational routines took shape during the making process as the artisans produced recognizable, repetitive patterns of interdependent actions carried out by multiple actors. An artifact-centered understanding of the design process reinforced a common misconception about routines: that they implicitly embody determinism.

During the local workshops, we all conceived of cultural artifacts as social and essentially non-commercial. They are linked to different concepts of time and space and, as I have previously discussed, they heralded social attributes based on collective inheritance and the actions generated around them. In this context, the we facilitators considered our mediating role as essential to the potential knowledge exchange between designers and artisans, and concluded that the sharing process achieved during workshops led to learning and promoting the empowerment of collective knowledge (Adams, 2014). There was an agreement between facilitators and artisans on the importance of full attention to the development of tools, processes, future workshops, post-reflections, and the transformation of prototypes (Blomberg and Karasti, 2013). It was also crucial that generated collective knowledge was spread to other women within the

Figure 4.8 Organizational routines during design workshops: Artisans Rosalía Bautista and Rosario Nicolás being observed by the oldest member (grandmother) sitting in a routine circle dynamic for activities during design. San Pablo Tijaltepec Design Workshop. CADA Foundation®, 2015.

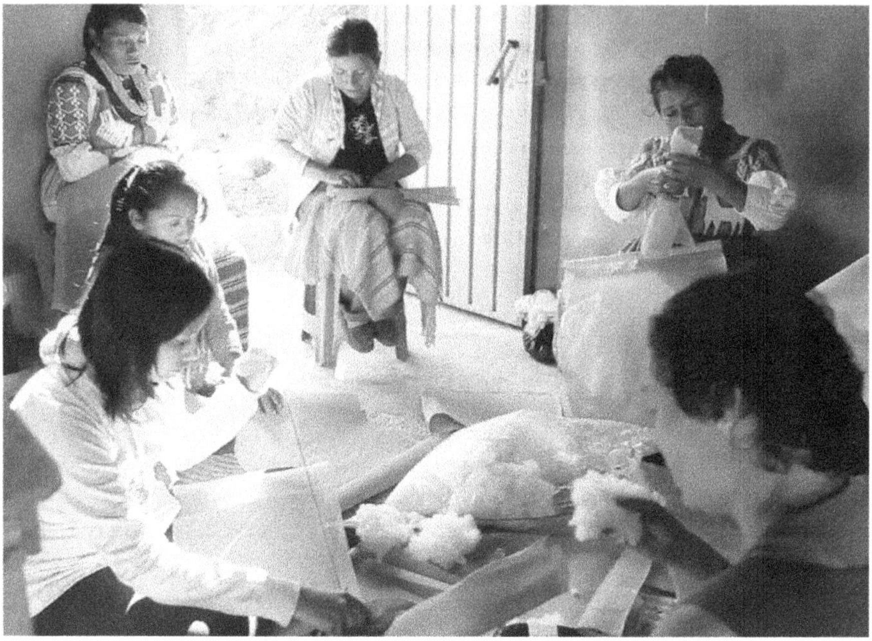

Figure 4.9 Experimenting together with new materials artisans had never used before. San Pablo Tijaltepec Design Workshop. CADA Foundation®, 2015.

communities, as naturally occurs with traditional methods of textile production, since women gather in each other's homes to embroider or spin communally.

In his landmark anthropological text, *How Societies Remember*, Connerton (1989) introduces English philosopher Michael Oakeshott's distinctions of moral forms, especially as they are applied in new and different contexts. He quotes Oaekshott's argument that "one is the reflexive application of a moral criterion and a kind of morality that is a habit of affectivity and behavior". This understanding of the application of moral criteria as "reflexive" rather than "principled" or "incorruptible" complicates the endeavor of an exchange. Yet a true exchange between different social and cultural contexts remains a necessary step toward confronting local and global transformations. Broadly, what we strove for through our process of approximation was to develop situations and parameters for more equitable work stages and practices, albeit adaptations based on cultures.

However, a moralistic intervention is not enough in many cases, as Papanek points out in his book, *Design for the Real World: Human Ecology and Social Change*. Papanek explains:

> Years of experience have convinced me that "instant experts" never work out. When foreign experts are brought to developing countries and faced

> with new problems, they frequently can provide answers that seem both sensible and workable. Their apparent ability to penetrate to the nub of the problem is illusory: unfamiliar as they are with the cultural background of the country, religious and social taboos, economic givens, and many other local considerations, they yet deliver an answer that appears convincing.
>
> [. . .] three weeks later the people on the ground quickly realize that, although they have solved the problem as stated, their "solution" has resulted in twenty or thirty new problems.
>
> (Papanek, 1971)

Thus, our challenge was to build a collaborative process that was neither moralizing nor blindly reductive. In sharing the process and analyzing the existing situation, we were aware of the need to balance our perceptions of traditional versus transitional elements (Bannon and Ehn, 2013; Simonsen et al., 2014). During our collaborative process, we maintained clear distinctions between the roles of involvement and non-involvement. We remained as reflective practitioners involved in the design and making processes. I also chose to remain impartial to any political disputes that may have arose during community affairs. Our interdisciplinary collaboration did not, at any moment, attempt to join items of knowledge about an already defined reality to create alternative design possibilities. Instead, our engagement was based only on generating an exchange of images and interest from different realities by constructing, composing and transforming several kinds of knowledge, social behaviors, and design practices (Gunn et al., 2013; Adams, 2014, pp. 20–24). The shared expertise of our multidisciplinary team ranged from graphic designers, to textiles designers, and to product designers, all of whom could draw upon years of practice in the field. The team was composed of designers with backgrounds in fundamentally different knowledge systems and who provided pertinent critical opinions throughout the work. This aspect was crucial to our collaborative work in order to establish dynamics for the workshops based on our deep reflections of our evolving experiences in the field. Overall, our meetings reflected a balanced decision-making process, and we resolved issues together through collective reflections and critical thinking (Bratteteig et al., 2013, pp. 117–144).

It was also essential that we facilitators were prepared to improvise, as we needed to quickly adjust our dynamics to respond to confrontational situations. Being open to tensions as they arose—and ready to explore difficulties and mistakes in the process and encourage group reflection—allowed for a deeper and more collaborative learning process. Indeed, despite our best efforts, many factors contributed to tension and discomfort within the group. Even though the participants assumed the sessions to be fruitful, not all of them had the same perspectives on specific aspects, nor did they all attribute the same values to the workshops. However, I ultimately found this diversity and richness in perspective to be powerful; supporting the artisan participants in situations to view through a different lens reduced their insecurities toward working beyond convention, thus revealing unpredictable possibilities.

Participation was best fostered when all the participants were included in the process and felt part of the common objective by helping define it and having a voice in the results (Blomberg and Karasti, 2013; Bratteteig et al., 2013). In the goal of generating a democratic and empowering work environment, we facilitators considered principles such as mutual respect for different knowledge forms and a commitment to openness during dialogue. All of this relied on our abilities to reflect upon our experiences with the artisans after each workshop, and we would adjust or transform the ensuing workshops in order to be most accommodating. Along with the artisans" use of traditional materials in new ways, and our introduction of new materials or methods, we facilitators observed and recorded workshop spatial considerations—that is, the use of space and the dynamics both materials and space generated on human interactions.

Throughout workshop transformations, we facilitators thought about the symbolic meaning of space and its impact throughout the design process, and observed the interactions between behavior and the physical environment, and the social identities of the artisans' communities in relation to their environments.

The ways in which all participants—including the design practitioners, artisans, and authorities (when they attended our meetings)—used open and public space changed significantly over the course of the project. From public squares to institutional buildings, bodies, and spaces engaged in open dialogue, and we witnessed different relationships between participants and their environments take hold. Our encounters were not confined to a given space or even a village; rather, we relocated often during our three years of work. The appropriation of spaces by our design process was causally related in an organic and fluid way to our relationships with the perimeters and objects.

John Zeisel (2006) describes how designers, in their creative processes, are very much like researchers, in that they conduct their work within well-established "environment-behaviors" to achieve greater design creativity and

Figure 4.10 Working environments: open public and private spaces. Initiation of workshops.

Left illustration; Santo Domingo Tonaltepec municipal open space. Right; working group entering artisan Juanita Bautista's house where the first series of design workshops would take place. San Pablo Tijaltepec. CADA Foundation®, March 2015.

research efficacy. The team and I were interested in the constant activity of all artisan participants and keenly observed and annotated their body language, which was particularly relevant when linguistic and cultural translation and interpretation were required. The human body is a rich resource for its ability to inform and inspire meaningful interactions as well as its ability to wander in search of physical location. Materiality provides great opportunities for studying our embodied selves. Throughout the process, we were interested in exploring the operations of power and resistance at the micro-level of the individual home and small community, analyzing modes of securitization and fortification utilized in the interest of wealth and power, and documenting the ways in which space and place are transformed by changing socio-economic and cultural demands.

Phase III: Prototyping, Artisanal Production and Market Exposure

After a deeply transformative Phase II, we facilitators moved on to begin a new stage of the project's workshops. This could only happen after our experiences with the artisans over the course of the previous phases, and after our team meetings where we reflected and adapted the project's design process and methodology to best fit the aims and goals of the artisans. The initial goal in this phase was to outline the artisans' prototypes made with materials that could eventually become detached from the traditional in inter-communal and interdisciplinary communities. These artifacts were the result of a design process focused on the collaboration between artisans and facilitators through a developing and deepening dialogue that was fostered over the course of the project. This involved discussions about the use of traditional techniques and its application to new and different materials.

Prototyping of the artisan's artifacts created out of new selected materials was studied as from both the point of view of the artifact's development and the viability of its commercialization with the hope that each artifact could be continually produced by the artisan communities. The artisan's prototypes also enabled our analysis of their constructive feasibility so that we could provide the artisans with timely feedback regarding potential iterative changes. These suggestions aimed, again, to provide support for the artisans' abilities to continue producing their artifacts beyond the duration of this project's workshops.

A process based on social design was implemented in a sequence of workshops based on a series of transformative practices which defined our course of action. We produced design situations to elicit the opportunity for artisans to transform their creations through organic workshop environments. Connecting with artisan traditions offered us facilitators insight into artisan crafts and their transmission of cultural and social values. The act of "making and unmaking" provided us a means by which we could support artisan decisions to form and transform new material artifacts for practical use and reflective of their cultures and histories. As workshops transformed, we witnessed and recorded these transformations, observing how artisans appropriated elements we could exchange with them. Transformations in relationships between facilitators and artisans were generated by collaborative dialogue and through our uses of space.

The very materialization of each artifact, both from its formal realization point to its functional 3D form, represented a significant transformation. The materialized cultural artifact showed whether or not the production process, the materials used, the way the prototype worked, its aesthetic, and its final effect agreed with the proposed initial conditions for its execution. This reflection process toward the conceived artifact and compared to the physical artifact was important during Phase III.

Instead of acting as agents of change, we considered the design process itself as a system with dynamics which followed a rhythm of flow, and which, through a complex network of connections, wielded the power to generate new forms of social relation, spatial, temporal, and material values. Within this design, opportunities for agency generated themselves in vital ways; however, the pursuit of a collective goal remained paramount. In keeping with the structure of the oral workshop and design workshop, graphic representation of knowledge was shared in an open context, allowing for participants to share and compare their worldviews and goals freely. The workshops went a step further to generating knowledge exchange and also transform new material into prototypes through traditional crafts. These processes of transformation allowed us facilitators to further understand the artifacts within their contexts. Thus, tacit knowledge and elements originating from various heritages and contexts were incorporated.

We focused on the semantic language of the artifacts within the two indigenous communities, which often carry complex symbolic meanings. In this vein, Ian Ewart (2013) proposed a concept of knowing by design. Nigel Cross analyses Ewart's "writing" concept:

> Continually doing the same thing makes it increasingly habitual and less likely to change. Designing does not therefore require a special designerly way of knowing, but depends instead on a designerly way of doing: design on-the-job, rather than in the mind [. . .] Design, in other words, does not exist per se, but only as a part of the performance of making.

Consequently, we focused on workshops through the making and unmaking of objects, generally de-emphasizing their final forms to give relevance to the collaborative making and unmaking processes that relied on traditional crafts. Artisans and social design practitioners all worked together to experiment with new materials or forms that either party had not experienced before. During the workshops, the artisans were free to take samples apart, cut, shape, and unravel them. These processes eventually would lead from prototyping to a defined and selected group of artifacts that communities could continue to produce and take to new markets venues. By this phase, the workshops were adapted sufficiently to establish a viable method in which the artisans could participate most effectively during the development of their artifacts while they also gained the opportunity to discuss, study, and eventually implement a marketing strategy for new exposure.

During this phase, artisans and facilitators considered physical properties such as texture, geometry (size, location within space, etc.), material (weight, flexibility, quality), and dynamic properties (or, how material artifacts can engage with multiple senses). The workshops were dependent on these explorations aiming to transform the artisans' worldviews toward tangible variations through contexts. We facilitators encouraged the artisans to take control and explore, and not to worry about the outcomes or feelings of failure. This process was built upon each of each facilitator's and artisan's background knowledge and sense of proportion, composition, and form. In our conversations with the artisans, they were noticeably confident in choosing colors and shapes. Ultimately, artisans and facilitators generated several new artifact forms through the use of new tools and materials, attempting to connect disparate design ideas while also staying rooted in practice and leaving space for discussion and mutual learning.

By this phase, we facilitators recognized a cycle where the artifact and production task are mutually interacting and reinforcing, both developing each other and defining themselves independently. As Tone Bratteteig writes in the Handbook of Participatory Design, "A task implicitly sets the requirements for the development of artifacts to support it; an artifact suggests possibilities and introduces constraints that often radically redefine the task for which the artifact was originally developed" (Bratteteig et al., 2013, p. 97). Later, she further defines the purpose of such a cycle: "[To] actively involve users in the analysis as active subjects, and the analysis becomes not only a joint activity of understanding the contextual conditions for the design, but also an activity of exploring opportunities for change." Indeed, the scenarios and interactions we facilitated were always unpredictable, only becoming clear in a very gradual manner over time. As Jesper Simonsen (2014) likewise believes, this natural cycle of gradual production is essential to the design process. All participants, facilitators, and artisans alike, understood that the task was defined by an original artifact that we began with at the start of each workshop session, and which was redefined during each session for each ensuing session. To reach this point in the project, and for us facilitators to recognize the importance of the artifact-task cycle, it was crucial that we had produced recordings and documented the entire workshop process through phases. The recorded data, or what I refer to as meta-research throughout Chapter 5, served as a basis for team reflection and our collective decisions to adapt the process over time and in consultation with the artisans.

Meta-research: Picturing Sounds and Movement of the Making and Unmaking

The video footage we facilitators produced portrays some of the different environments within which the workshops allowed for collaboration between facilitators and artisans. By watching excerpts of the process, we may gain a

Figure 4.11 Some samples of the resulting prototypes. Santo Domingo Tonaltepec. CADA Foundation®, 2016. The resulting prototypes display a weaving-together of analysis and design. An interactive, hermeneutic process known as the "task-artifact cycle" (Carroll et al., 1991).

more holistic sense of our collaborative practice throughout the project. My intention is to show how the material gains in form and becomes an artifact object, show its identity, reveal the context of its use, show its material quality, and display how the body reacts to it. It is central to the purpose of this project to emphasize the collaborative design process that artisans and facilitators engaged in to create the artifacts.

As Jay Ruby (2000) pointed out, the aesthetics of figurative images also have a political potential; not necessarily thematically, but in their effects and ability to trigger visual memories. To Bal, their political efficacy lies in their ability to make meaning on the different grounds of any semiotic system. In the editing process, the team and I tried to focus on those details that are important to the meaning of the project yet would not be obvious to an unfamiliar spectator unless pointed out. We particularly explored how sound and movement operate in conjunction with person and material in the search for the material's identity. Sound and movements are always expressive, as they bring together internal and external elements.

Moreover, sound and movement are indispensable as ways to understand meaning, for they are fundamentally affect-based. Bal defines sounds as "nothing but movement" (Jay Ruby, 2000, p. 83). This approach allowed me to escape the trap of the viewer identifying with "the other people"—referring to those represented—thus canceling the difference between self and the other. Ruby (2000) warns against this trap, and Dominick LaCapra (2001) coined the term "empathic unsettlement" to describe it. This danger was constant in my mind and in the minds of my fellow practitioners throughout the process. In considering LaCapra's concept, we tried to curate a selection of excerpts that maintain awareness of a "distinction between one's own perceptions and the experience of the other" (Bennett, 2010, p. 8). The narrative must be shaped by the "how," instead of the "who" or "what," since it must be dominated by the form and its transformation thus rendering useless questions of voice, authority, and authorship.

This narrative manifests itself not only through narration, but also through camera and audio techniques emphasizing certain aspects of the participants and their materials. Jay Ruby (2000) argues that the desire to entertain can eclipse the original intention of informing an audience. Thus, cooperative,

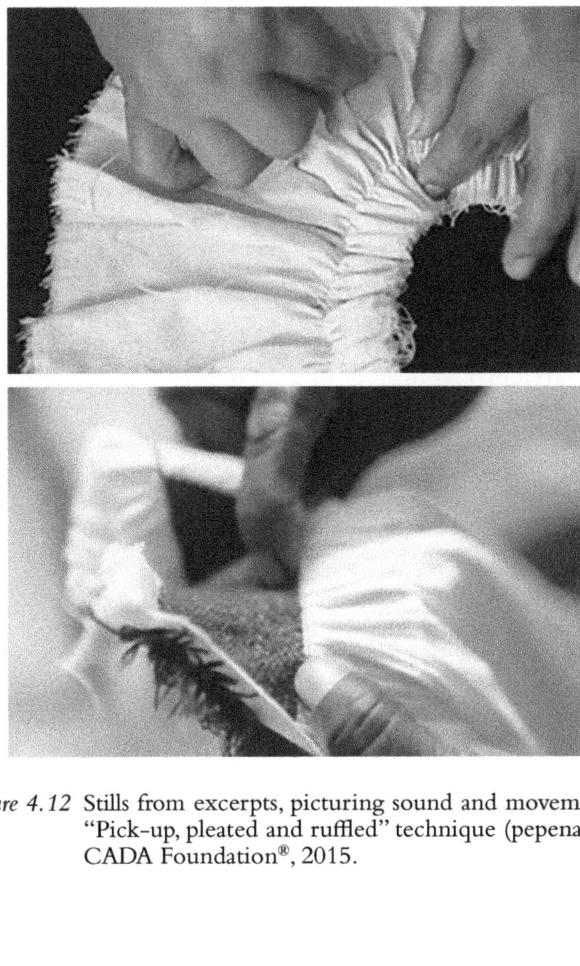

Figure 4.12 Stills from excerpts, picturing sound and movement through the material. "Pick-up, pleated and ruffled" technique (pepenado). San Pablo Tijaltepec. CADA Foundation®, 2015.

collaborative, and subject-generated films are a response to the need for a rethinking of authorship. As Ruby notes, "[T]he audience's role construction of meaning in (ethnographic) film is a complex one ... On the other hand, there are the larger questions concerning the role of the reader/viewer in the construction of meaning" (Ruby, 2000, p. 181). *Meta-research: Such example of exploration was partially edited online and can be viewed in Vimeo: https://vimeo.com/490534957/94ca57693f.

Transformative Practice and the Collective Choice

In this chapter, I introduce perspectives that helped us facilitators to understand our challenge as social designers in building a collaborative process that was neither moralizing nor blindly reductive. As such, it was necessary to maintain the social and cultural context where the practices took place, and to leave room for heritage as the enterprise of knowledge. The practice itself involved a form of reflective communication for social design, and developed collaborative interactions framed by the process of making things together among people with different cultural backgrounds, economic systems, and core beliefs. I characterize this practice through its potential for transformation at the social and economic levels. As I have previously mentioned, we facilitators were not looking to solve any single design problem in the artisan's local context, but we instead sought to support the artisans' and their communities' goals to connect with a global network while also maintaining their historical and cultural identities. We structured transformative situations during the design process that I term proximities of design.

Transformative situations were indeed the product of a system of exchanges through which the making of artifacts became a shared experience. We generated ways of sharing knowledge and answered some questions about whether or not a practice of social design can promote scenarios that strengthen communities. Overall, we facilitators developed workshops that supported local communities in their efforts to gain agency within global markets, while also opening new research avenues between collaborative work and social design practices within the field of material culture.

Notes

1 Under Sistemas Normativos Indígenas, Municipal authority—called Sistema de Cargos o Escalafón—is held by two individual roles: the municipal president, and the commissary for common property, who leads the agricultural community, as well as commissions and committees who ensure all activities and need of the community are met.

2 "Güera," meaning "a blond-haired person," is common Mexican parlance for any foreign-looking white woman.

3 Refer to Chapter 2 for a discussion of the proximities of design concept, and to Chapter 5 for our facilitator applications of this concept during the workshop practice.

4 El Camino de los Altos—ECLA—is a non-profit association established in México as a result of a twelve-year collaboration between 120 indigenous Mayan weavers from five different municipalities in Los Altos de Chiapas, and French non-profit association El Camino. El Camino is a group of eight professional designers, working together to create a premium-quality textile collection. Among these designers was Madalenna Forcella, along with founder Veronique Tesseraud. The weavers of ECLA live and work in the Chamula, San Andres Larrainzar, Pantelho, Oxchuc, and Zinacantan municipalities spread throughout the mountains surrounding San Cristóbal de Las Casas, close to the Guatemalan border and in an economically underprivileged area.

5 Anthropologist Luis Joel Morales Escobedo (Luis Morales) works in the recognition of Craft Altos de Chiapas Region, and lives in San Cristóbal de las Casas, Chiapas, where he served as Connecting Area Chief at CECATI No. 133 San Cristóbal de las Casas (Chiapas) Work Training Center (Centro de Capacitacion para el trabajo), training workers at several of the Chiapas state communities since the 1970's. He founded Colectivo and acted as a consultant in the CECATI No. 166 Documentation and Productive Process Case de Study. He is the founder and president of Kun K'un Talleres de Artes y Oficios in Workshops of Arts and Crafts in 1995 in San Cristóbal de las Casas, which promotes the use of natural dyes in textile and Ceramics. He also organized artisan groups in the region known as Los Altos de Chiapas jointly with Maddalena Forcella, Gertrude Garman, Mayte Bellato, Manuel Núñez, and Lucía Méndez.

6 Omar Fernández, Architect and Teacher Universidad Iberoamericana at Puebla y la Universidad Mesoamericana, in Oaxaca. He trained at Universidad Iberoamericana, Mexico City. He designs and builds architecture, interiors and furniture projects, and participates in several social programs like Santa María Yavesía's furniture workshop and the LAB U.C. group. As a member of the restoration team at the former Convento de Santo Domingo de Guzman in Oaxaca, and the restoration of the Capilla Abierta in Teposcolula, the structural restoration of the Templo de Santo Domingo de Guzman in Yanhuitlan he had extensive experience working with craftsmen, traditional techniques and contemporary uses and needs.

7 Weaver Eric Chávez is a fourth-generation textile artisan. He wove his first carpet as a young 12-year old and encouraged his parents to work with natural dyes. He is a university graduate and worked at the Museo Textil de Oaxaca (M.T.O) as an education director, coordinating different social collaboration programs, involving designers, cooperatives, and groups of textile craftsmen in their Oaxacan communities. He set other initiatives, workshops, and exhibitions of textile craftsmanship, in which guest master weavers and textile craftsmen from Oaxaca, other Mexican regions, or foreign countries are invited to share their techniques.

8 Ludovica Morales is a young graphic artist from Oaxaca, with tzetzal cultural and language traditions, who collaborated as facilitator in the children workshops. She was raised in San Cristóbal de Las Casas where she finished her primary education. In 2004 she collaborated with her artwork in the book Losha, Editorial Fray Bartolomé de Las Casas, PACMYC, CONACULTA.

9 The "intangible cultural heritage" means the practices, representations, expressions, knowledge, skills—as well as the instruments, objects, artifacts and cultural spaces associated therewith—that communities, groups and, in some cases, individuals recognize as part of their cultural heritage. This intangible cultural heritage, transmitted from generation to generation, is constantly recreated by communities and groups

in response to their environment, their interaction with nature and their history, and provides them with a sense of identity and continuity, thus promoting respect for cultural diversity and human creativity. (Source: UNESCO Convention for the Safeguarding of the Intangible Cultural Heritage (ICH) 2003, Global https://unevoc.unesco.org/home/TVETipedia+Glossary/lang=en/filt=all/id=723).

10 Taller de Leñateros. http://www.tallerlenateros.com/gaceta_web/eng/gazette.htm Publishing collective operated by contemporary Mayan artists in Chiapas. Founded in 1975 by poet Ambar Past, Taller has published the first books written, illustrated, printed, and bounded (on paper of their own making) by the Mayan people in more than 400 years. Among its multiple objectives it is worth mentioning their aim to document, advance, and promote the native and popular cultural values: the native languages literature, the visual arts, and the painted codex. Leñateros work on rescuing old and near extinct techniques such as the extraction of dyes from wild herbs, and the recovery of native languages.

11 Sociedad Indígena de Chiapas. http://culturadelosindiosmayas.mex.tl/frameset.php?url=/intro.html

Sna Jtzi´bajom "Cultura de los Indios Mayas A.C." Founded in 1982 under the name of Sociedad Indígena de Chiapas (Native Society of Chiapas) and legally constituted on May 23 1983, thanks to the initiative of pioneering Tzotziles and Tzeltales translators, and important figures like Rosario Castellanos, the poet Jaime Sabines, linguist Doctor Roberto Laughlin from the Smithsonian Institute, and several national and international anthropologists, linguists, Institutions and Foundations. The Sociedad has supported different aspects and moments of the work to preserve control over their own cultures and achieve a creative state of peace and order, with security and food sovereignty, with justice and dignity as established in the 169 OIT agreement.

References

Adams, B. 2014. Crafting capacities. In: Yelavich, S. and Adams, B., eds. *Design as future-making*. New York: Bloomsbury Academic, pp. 20–24.

Arellanes Meixueiro, A. and Simón Reyes, A. 2011. Pasado y Presente de la Alfarería en Oaxaca. Oaxaca: Instituto Tecnológico de Oaxaca.

Balaram, S. 1995. Product symbolism of Gandhi and its connection with Indian mythology. In: Margolin, V. and Buchanan, R. eds. *The idea of design*. Cambridge (Massachusetts), London: The MIT Press, pp. 127–143.

Bannon, L. J. and Ehn, P. 2013. Design. Design matters in participatory design. In: Simonsen, J. and Robertson, T. eds. *Routledge international handbook of participatory design*. New York and London: Routledge, pp. 37–63. 32.

Bennett, J. 2010. *Vibrant matter: a political ecology of things*. Durham: Duke University Press.

Blomberg, J. and Karasti, H. 2013. Ethnography. Positioning ethnography within participatory design. In: Simonsen, J. and Robertson, T. eds. *Routledge international handbook of participatory design*. New York and London: Routledge, pp. 86–116.

Bonfil, G. 1987. México Profundo. Una civilización negada. México, DF.: Grijalbo.

Bratteteig, T., Bodker, K., Ditrich, Y., Holst Mogensen, P. and Simonsen, J. 2013. Methods. Organising principles and general guidelines for participatory design projects. In: Simonsen, J. and Robertson, T. eds. *Routledge international handbook of participatory design*. New York and London: Routledge, pp. 117–144.

Carroll, J. M., Kellogg, W. A. and Rosson, M. B. 1991. The task-artifact cycle. In: Carroll, J. M. ed. *Designing interaction: psychology at the human-computer interface.* New York: Cambridge University Press, pp. 74–102.

Connerton, P. 1989. *How societies remember.* Cambridge: Cambridge University Press.

Creswell, J. 2013. *Qualitative inquiry and research design: Choosing among five approaches.* 3rd edition. University of Nebraska, Lincoln: SAGE, pp. 94–95.

De Avila, A. 2013. Hilo del País de las Nubes. Oaxaca: Museo Textil de Oaxaca.

Ewart, I. J. 2013. Designing by doing: building bridges in the highlands of Borneo. In: Gunn, W., Otto, T. and Smith, R. C. eds. *Design anthropology: theory and practice.* London and New York: Bloomsbury, pp. 85–99.

Geertz, C. 1973. *The interpretation of cultures: selected essays.* Vol. 2019. Basic Books.

Geertz, C. 1983. *Local knowledge: further essays in interpretive anthropology.* Vol. 5110. Basic Books.

Gossen, G. 1974. *Chamulas in the world of the sun: Time and space in a Maya oral tradition.* Cambridge, Massachusetts: Harvard University Press.

Gunn, W., Otto, T. and Smith, R. C. 2013. *Design anthropology: theory and practice.* London and New York, Bloomsbury.

Hernández Sánchez, G. 2013. *Ceramics and the Spanish conquest.* Leiden: Brill.

Kuchler, S. 2011. Materials and design. In: Clarke, A. J. ed. *Design anthropology: object culture in the 21st century.* Viena: Springer, pp. 130–141.

Lacapra, D. 2016. La historia y sus límites. Editions Bellaterra.

Lazcano, J. 2005. El trabajo artesanal.Insatituto Naciona de las Mujeres, México

Makovicky, N. 2011. Erotic needlework: vernacular designs on the 21st century market. In: Clarke, A.J. ed. *Design anthropology: object culture in the 21st century.* Viena: Springer, pp. 155–168.

Margolin, V. 2002. *The politics of the artificial.* Chicago, London: The University of Chicago Press.

Mindling, E. 2011. Barro y Fuego. El arte de la Alfarería en Oaxaca. México D.F.: Editorial Arte Oaxaca, p. 70.

Novelo, V. 2003. La capacitación de artesanos en México, una revisión. México D.F.: Plaza y Valdés.

Ornelas, Y. and Gregory, J. 2011. Cultural artifacts as emotional catalysts and social mediators in design research. 4th Conference on Design Research. Netherlands: ITT Institute of Design.

Ortiz Escamilla, R. 2012. Las Rutas de la Tierra del Sol. Oaxaca: Universidad Tecnológica de la Mixteca.

Papanek, V. 1971. *Design for the real world: human ecology and social change.* 2nd ed. Chicago: Academy Chicago Publishers.

Pellicer, J. 1995. Artesanos del Porvenir. México, D.F.: Secretaría de Educación Pública, Universidad Autónoma Metropolitana.

Proada. 2002. Manual Programa de Apoyo al Diseño Artesanal. Conaculta.

Ranjan, M. P. 2011. Design for good governance: information and communication as a platform for change. Keynote Lecture at the 4th International Conference on Contemporary Computing, Jaypee Institute of Information Technology, 9 August 2011, Noida.

Recchia, F. 2011. Radical territories of affection: the art of being stalker. In: De Cauter, L., De Roo, R. and Vanhaesebrouck, K. eds. *Art and activism in the age of globalization.* Rotterdam: NAi Publishers, pp. 160–171.

Ruby, J. 2000. *Picturing Culture: Explorations of Film & Anthropology*. University of Chicago Press.

Simonsen, J., Svabo, C., Malou Strandvad, S., Samson, K., Hertzum, M. and Hansen, O. E. 2014. *Situated design methods*. Cambridge, London: The MIT Press.

Stallybrass, P. 1996. *Subject and objects in Renaissance culture*. Cambridge University Press.

Telier, A., Binder, T., De Michelis, G., Ehn, P., Jacussi, G., Linde, P. and Wagner, I. 2011. *Design things*. Cambridge, MA: MIT Press.

Tunstall, E. 2013. Decolonizing design innovation: design anthropology, critical anthropology, and indigenous knowledge. In: Gunn, W., Otto, T. and Smith, R. C. eds. *Design anthropology: theory and practice*. London and New York: Bloomsbury, pp. 232–250.

Turok, M. 1988. Cómo acercarse a la artesanía. México D.F.: Editorial Plaza y Valdés.

UNESCO Convention for the Safeguarding of the Intangible Cultural Heritage (ICH) 2003, Global https://unevoc.unesco.org/home/TVETipedia+Glossary/lang=en/filt=all/id=723

Zeisel, J. 2006. Inquiry by design: environment, behaviour, neuroscience in architecture, interiors, landscapes and planning. New York, USA: W. W. Norton.

5 Field Notes

Interviews, Workshops, and Prototypes

In this chapter, I curate a set of resources that represent an explanatory presentation in the form of a catalog or dossier. This section includes my documentation of the design process and prototypes, as well as about the context, aims, and findings surrounding the artifacts. I also offer brief accounts of the situations which properly encapsulate the "proximities of design" concept and which were key to the design process. Those situations were undertaken with respect to people, places, and materials. At the same time, and through my reflections on those situations, I suggest some limits to the designer's role and discuss the implications of those restrictions for a critical design practice. In this section, I also expand upon the ethnographical research conducted during field work to meet activists, anthropologists, artists, and educators who possess extensive experience working with artisan communities in Oaxaca and Chiapas.

The structure of this chapter helps to situate my project practice more clearly within a context of scholarly research. This is in part due to my aim to articulate a critical reflection that will persist beyond the duration and location of any physical exhibition of this design process. Technique and material exchange among participants were a significant aspect of these reflections. For us design practitioners, the traditional academic reporting method seems an insufficient format to present our research. While we are able to talk about the process in order to control and expand the dissemination of our work, writing about the work never seems to do the process sufficient justice. I believe that the reader will gain a more holistic sense of the practice through examining some images of the process, through reading the researchers' and artisans' reflections on the project, and by beholding the artifacts. By developing new forms of communication about design by "eliciting situations" this way, we may be able to enhance the potential for new transformations of design.

I provide the information in this chapter in such a way to more effectively structure the project's events and to offer the reader a greater understanding of this project due to the fact that we facilitators and artisans developed this research episodically over an extended period between 2013 and 2017. Within this timeframe, the research and workshops transformed from general to specific as we developed our project's workshop to recognize the more concrete needs of the participating communities. Moving forward, I organize the

DOI: 10.4324/9781003360803-7

information for the book around two cores: the first consists of exploratory research during travel, visits, and interviews and will be described throughout the following sections; the second consists of detailed accounts of the workshops and facilitator's and artisans' meetings, and will be described as case studies.

Documenting Processes and Outcomes: Meta-Research

In my own practice, I aimed to substantively contextualize processes, and I focused on the "who" and "how" of "*el hacer*"—"the making." Based on my prior design knowledge, I created a series of guide notes describing the progression of design experiences. Based upon my experiences in the field, I do not necessarily fashion these notes in a linear or chronological order. Instead, my notes usually serve as explanatory texts that contextualize the resources, stories, events, and sites (see Figure 5.1). The texts also may include graphic material such as visual landmarks to trigger concepts and memories with the intention of being exhibited. Based on a chronological narrative dictated by events supported on my knowledge of design process, I structure diagrams eliciting what I called "proximities of design"—or, transformative situations during the design process.

From a design-practice point of view, I struggled to organize the material I gathered in such a way to exemplify how it helped me conduct my research and share it in this book. In the end, I organized information related to content and objects into several case studies so that I could present and describe the main characteristics and locations of the process and offer a sounder basis for the readers' judgment. The fundamental structure spans across variety of

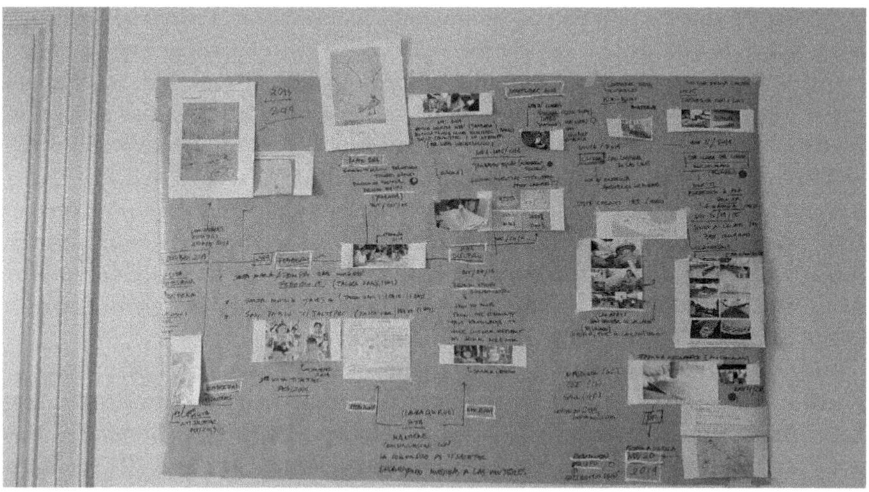

Figure 5.1 Notes on field trips and meetings. Process work. Barcelona. CADA Foundation®, 2017.

settings—both physical environments and social-organizational settings—from long before the workshops, to during the process, and long after. I evaluated the different cultural expressions contained within these spaces by understanding each expression in relation to their own context, and in reference to Bonfil Batalla's work that I discuss in Chapter 2.

Exploratory Research Travel and Visits (2013–2014)

Between 2013 and 2014, I began conducting field research which required my travel across México to meet and interview community members, designers, anthropologists, and master craftspeople who had been involved in CECATI government crafts training centers or with community-artisan development programs. My visits to various artisan communities were aimed to hear experiences and learn the practices of communities in their entirety, or of specific and relevant persons, in order to observe local circumstances, make contacts, and evaluate potential synergies. These visits were also solicited by the community members themselves, who sought the support for market exposure. I traveled using public transportation over long distances which was both time and energy consuming, but my experiences during that time allowed to gather firsthand impressions about the country, its political issues, and its people during their daily activities and in different spaces.

My travel brought my attention toward many aspects of relevant contemporary political, economic, and social issues. The México I observed was a vast and complex society containing a heterogeneity of cultures. As a foreigner, I maintained a discreet approach to political discussions and other social conflicts that I perceived as affecting both rural and urban communities. Sometimes, I noticed graffiti and other graphic expressions as manifestations of popular protests about topics that mattered to people in their daily lives. Many of these tangible expressions provided me greater insight into local perspectives toward cultural tensions, including appropriation and many other social realities. I consider the effects of such realities—poverty, public access to opportunities, drug and human trafficking, and migration—as part of my observations, and I attempted to record my experiences systematically. I used video and photographic tools in order to document the wider perspectives on these local realities and circumstances. This effort also provided me a more vivid impression of the sites I visited for my recollection more than mere writing and note-taking could allow. I describe my experiences during these visits below.

2013—February. Site Visit to San Marcos Tlapazola (Tlacoula de Matamoros, Oaxaca)

On February 8, 2013, I traveled to San Marcos Tlapazola with Fátima Díaz and the photographer Paul Milajos. We met Santiago artisan family. The aim of the visit was to analyze commercial logistics, observe trade processes and techniques, and study accessibility and marginality indexes. These communities sought more direct market exposure and were happy to share their processes with us in exchange for our support.

The towns in the state of Oaxaca are representative of the Zapotec population as well as the crafts they make. The town of San Marcos Tlapazola, for example, is famous for the red clay pottery its artisans produce. It is a small rural community located approximately one hour from Oaxaca City and roughly ten minutes south of Tlacolula de Matamoros.[1] The clay of this town is particularly unique due to its terracotta style with different intensities of red color. There are about 300 potters in the village, but they work sporadically. The crafters make each piece of pottery by hand. Each piece they produce is of exceptional quality and demonstrates their great craftsmanship. Women take care of every step in the production process. Each artisan mixes their clay, shapes their piece, and bakes their artifact. These artisan women protect a very old tradition that is important to their community. Since the late 1980s, however, the village potters have been experimenting with new forms and designs which has bolstered sales and recognition for the village.

Due to its proximity to the city of Oaxaca, this community seemed the most exposed to wider markets. In the surrounding landscape, I observed the threat of deforestation due to the community's continuous burning of firewood to cook their mud. Within the community, the artisans showed an openness reflecting their continuous exposure to foreigners and the artisans' history of market exposure.

2013—February. Site Visit to San Martin Tilcajete (Ocotlan, Oaxaca)

On February 15, 2013, I traveled to San Martin Tilcajete with Fátima Díaz. We met with the artisans Florencio Sánchez and Paula Sánchez. The aim of this visit, similar to the previous one, was to analyze commercial logistics, observe trade processes and techniques, and study the community's accessibility and marginality index.

San Martín de Tilcajete is about 30 kilometers from the city of Oaxaca. It is the town where *alebrijes* are born: representations of imaginary animals carved in wood with a combination of bright colors. The meaning of *alebrije* comes from the three words "joy" for colors, "witch" for mysticism, and "*Embije*" which means to dye or paint something with red. The artisan knowledge for elaborating with wood carvings is passed down generationally. First, a piece of copal wood is chosen to be cut with the machete and then with a knife and blade. Next, the wood is dried and cured to prevent it from being damaged by moisture or insects. Then, it is carved little by little, as the artisan anticipates the alebrije's form. Finally, it is carefully painted using vivid colors including red, blue, pink, green to add additional graphical designs unique to each piece.

Both production and sale of these artifacts happen within the artisans' home; they live among stacks of copal logs, with their tools to carve and paint. Thus, children and young people learn this trade from the oldest members of their family and community. The inhabitants of Tilcajete are born and grow closely linked to the elaboration process as part of the family environment.

San Martín Tilcajete is another community that shows existing ties to the broader market and international trade. The community is important to Oaxaca, and their artifacts are commonly selected as souvenirs among tourists. Artisans and the broader community also show openness to collaborating with outsiders and are used to participating in exhibitions. In the 1980s, the community benefitted when Oaxacan handicrafts were promoted to broader audiences. The carvings from Tilcajete received particular interest which ultimately increased sales and the further development of their techniques and morphology through their crafts. Currently, the elaboration of these crafts is the base of the economy in Tilcajete.

2013—February. Site Visit to San Pedro Cajonos (Villa Alta, Oaxaca)

On February 22, 2013, I traveled to San Pedro Cajonos with Fátima. We met with the artisans of the Martinez Family from an atelier called the *Wed Do Sed Association*. They produce and work with Creole Silk; from silkworm farming, to silk thread production, and also silk weaving.

San Pedro Cajonos belongs to the district of Villa Alta[2] within the northern sierra region. Its population, mainly Zapotec, is known for its original indigenous crafts. San Pedro is one of the towns that make up the silk road trilogy in the municipality of San Francisco Cajonos. There, we visited a group of artisans dedicated to the production of silk shawls, dresses, and jewelry. The artisan group was represented by the artist Moisés Martínez who, along with his brother José Martínez Velasco, has formed the Wen Do Sed Productores de Seda group to raise their own silkworms, both wild and cultivated.

The biggest challenge for the Martínez family, and for all silk producers in the Cajonos region, is to find the appropriate markets for the sale of their products. The process—from rearing the worms to coloring the pieces with natural dyes—takes months of work which is justly reflected by their prices.

There are two types of silkworms typically used to produce silk products: Creole silkworms and "improved" silkworms, and the latter are not born directly from the butterfly. There are subtle differences in color and touch between the silks produced with both types; the silk of Creole worms is softer and displays a more irregular color. The worm breeding process begins once the worms form. Once the worms are formed, rearing work involves keeping them well fed with leaves from the mulberry tree. After three weeks, the worms stop eating the leaves and change their color to a transparent green. That is the signal that indicates that the worms are ready to form their cocoons with the silk thread. The breeder then accommodates the worms on oak leaves and waits for the worms to do their homework.[3]

The spinning process begins when the cocoons of the worms are boiled to form a yarn. During the boiling and washing phase, the Creole worm cocoons

Figure 5.2 José Martínez spinning and weaving to obtain Creole Silk. San Pedro Cajonos. CADA Foundation®, 2013.

lose their yellow color and turn white. The spinning process is completed manually with a winch or a mechanical pedal device (see Figure 5.2). During this process, the balls of silk thread are formed and are then ready to go to the weaver. Once the silk is weaved and embossed, the pieces are ready for coloring with natural dyes. Artisans use *pericón*, Brazilian wood, indigo, and *cochineal grana*[4] as the base to obtain tonalities.

It is interesting to observe the whole process of the silk thread. It is a holistically structured environment and silk product manufacturing is conducted in an organic way. For some years now in this area, this ancestral technique of silkworm rearing, weaving, and garment dying have been recovering. While the community seems to be more isolated compared to the previous communities that I have visited, its members are equally friendly and open to collaboration with foreigners as those were prior.

2013—February. Site Visit to San Bartolo Coyotepec (Oaxaca Center District)

Also in February 2013, I traveled to San Bartolo Coyotepec with Fátima. The municipality of San Bartolo Coyotepec is located in the Central district of the Valles Centrales region, in the Mexican state of Oaxaca. The black clay crafts of Coyotepec come immediately to mind when considering the pottery of Oaxaca. It is one of the most famous pottery towns in México. It is also the largest in Oaxaca, with some 700 households actively producing. There, we met with the artisan Armando Magno.

The black color of the clay is due to its physical properties and not due to added colorants. The mud used to form clay in this pottery technique is highly sought-after. First, the mud is extracted from the soil and is cleaned to protect against impurities. This process can take up to a month to soak and set the mud apart from the rest of the soil. After this cleaning process, each pottery piece can take nearly 20 days to finish.

Once cleaned, the clay is traditionally molded into plates and then balanced on top of rocks so that it can be worked by hand. Artisans form large pieces—such as pitchers—from the base up; they add more and more clay to grow each piece. After the pieces take their larger shape, they are placed to dry in an insulated room to protect them from changes in temperature. The drying process can take up to three weeks. If the piece is to be polished, it is done when the piece is almost dry to achieve a glossy black finish. Artisans slightly moistened the surface of each piece and then rub it with a curved quartz stone. This compresses the surface of the clay and creates the metallic shine and dark color during baking. The pieces are then fired in holes under the ground or in ovens fueled by woodfire. Once complete, the polished pieces are a glossy black color and those that were not polished have a matte gray finish.

The community is already exposed to wider market processes. About 95% of the population participates in pottery production. Thanks to its robust sales, most potters make a comfortable living from their craft. Countless stores in Coyotepec sell locally produced pottery. It is possible to visit the artisan market and find the pottery in the center of town, next to a large tree, as well as the market at the entrance of the town. While the current production of pottery has lost most of its utility value, the products remain highly decorative and possess craft recognition worldwide.

2013—February. Site Visit to San Pablo Tijaltepec (Tlaxiaco, Oaxaca)

I traveled for the first time to San Pablo Tijaltepec with Fátima Díaz in February 2013. Local organizations working with this community informed me that artisans there had requested support to reach wider market exposure for their crafts. We also met with Juanita Bautista, a local artisan from San Pablo Tijaltepec, through the liaison and facilitator Laura Quiroz. Artisan women within the community produce embroidered textiles through pickup weave and gathered pleats using a vegetable fiber or palm.

San Pablo Tijaltepec is located in the Mixtec region in the state of Oaxaca. Its territory consists mainly of hills. It is found in coniferous forests and rivers. Unfortunately, this natural environment is negatively affected by a large-scale exploitation of resources, and especially the unregulated felling of oak and ocote trees for domestic use and construction. In 2015, according to the National Population Council, the municipality was classified with a very high degree of marginalization contributing to a large number of its male population's emigration to other places in search of employment. Women remain in the village to care for their children and domestic animals. They dedicate themselves to the activities of the field, the home, and the process of their craft work to produce garments for their personal use or for sale in local markets.

The canvas artisans use for each garment is a blanket displaying natural color, and the threads used for the embroidery are usually an acrylic yarn purchased at the local market. The embroidery technique of the blouse is known as gathered pepenado, because each row of basting is embroidered separately and then the threads are pulled together at the end of the design to give way to the representation of deer, birds, and elements of nature in geometric forms. The deer embroidery is inspired by a sacred figure from the Mixtec and is the most commonly embroidered element by the women of San Pablo Tijaltepec. Textiles are created by women under the precept of a deep sense of communal process. This is to say that they are collective patterns and, as such, must be respected. Girls are taught from an early age to embroider, and later to make their own garments.

The village has very difficult access to main roads, which are not properly maintained by the state. The roads that are accessible are also in very bad condition and are poorly connected to other important locations. This means the community faces a challenge connecting with other communities and markets. The community still maintains its original language—one of many Mixtec variations still spoken today—and the majority of community members do not speak Spanish. We experienced some difficulty communicating, although Juanita assisted to facilitate a dialogue between facilitators, artisans, and local authorities.

2013—February. Site Visit to San Agustín Etla (Etla, Oaxaca) Taller Arte Papel (Paper Art Factory)

I visited the Taller Arte Papel (Paper Art Factory) in San Agustín Etla in 2013. This workshop space, located in a former hydroelectric plant, uses native plants to create paper for kites, bags, jewelry, and more. San Agustín Etla, nestled in the mountains, is a town located about 30 minutes from Oaxaca. It is surrounded by mountains covered by forests as well as springs. Its landscape has served as a source of inspiration for many artists, painters, and craftsmen over the years.

The old Vista Hermosa hydroelectric plant stands at 1,700 meters above sea level in San Agustín Etla, Oaxaca. It is the current headquarters of the Taller Arte Papel Vista Hermosa cooperative, or better known as the Etla paper factory since its foundation. The cooperative is an initiative of the former plastic artist, patron, and cultural provocateur Francisco Toledo. The organization currently serves as a community project for artistic collaboration, and is both ecologically friendly and self-sustaining. Unlike others of the same type, the Etla Paper Cooperative produces sheets of handmade paper of various types and thicknesses by extracting natural fibers from trees, plants, barks, and flowers found both locally and in other parts of the world. Resources are extracted with sufficient knowledge of the local flora to avoid breaking the environmental balance of the community.

To produce the paper, artisans cook each fiber for two to three hours in order to soften the material, and then they are ground onto trays. Colors are added during the grinding stage to obtain a uniform pulp, which is then taken to racks to form leaves. Nopal slime is also added to make the paper more flexible. To give the paper color, artisans use *cochineal grana*, aquatic plants, achiote fruit, pomegranate, and indigo blue. The production of paper is carried out in a traditional way with instruments such as a Dutch pile, pots, buckets, racks, felts, a press, and clotheslines. The water necessary for the alchemy is also reused to contribute to the ecological and self-sustaining characteristics of the process.

The workshop specializes in the production of four or five types of paper that can be used for creative purposes. Notepads, sketchbooks, letterhead, envelopes, artist boxes, and stationery are also made. The importance of learning these techniques relates to the community's need to create its own source of income. Eventually, I sourced paper from this factory to record oral narratives during this project's workshops.

2014—February. Travel and Visits to Chiapas and Communities in Oaxaca. Interview with Ambar Past

On February 6, 2014, I traveled to San Cristóbal de las Casas, Chiapas, with Maddalena after first traveling to Tuxtla Gutierrez. My aim during this trip was to gather firsthand data for the project. I hoped to critically engage with the evidence I found there by drawing upon the existing literature and my discussions with other researchers and practitioners whose work had been focused in the context of Latin America. In particular, I hoped to further enrich my knowledge of widespread economic and cultural contexts across regions of México to gain a greater perspective on cultural values and identities. On February 7, we contacted poet Ambar Past for an interview.[5]

During our interview, Ambar explained her work and role as an activist in civil associations dedicated to promoting art as a cultural expression. Her work included theatre activities to strengthen local identity and vision, enrich the old and new cultural values, and broadcast and strengthen the indigenous cultural values of Altos de Chiapas through the anthology of oral tradition. Her main focus was to highlight local identity and safeguard it against foreign influence while viewing culture as static and intact.

For Past, an important aspect of her activism is to promote teachings of reading and writing in Tzotzil and Tzeltal languages among the indigenous children and youths of the Altos de Chiapas in order to strengthen their cultural identity. This also includes the dissemination of information about environmental, cultural, and educative values through radio programs for the benefit of local communities. Part of these associations' role includes preserving, documenting, and broadcasting the cultural traditions of the Mayan people. When I spoke with Past about my project, she aligned with the theories of Turok and shifted from unwelcoming to accusatory. An invitation to eat with her at her house ended with a tense conversation. Maddalena, who had introduced me to Past, defended the legitimacy of our approach as unrelated to colonialism and to any intentions to exploit the local communities.

2014—February. Travel to Chiapas/San Cristóbal de las Casas. Interview with Luis Morales

I also met and interviewed Luis Morales, the then director at CECATI 133. Morales is an anthropologist who had cofounded K'un K'un—a local Arts and Crafts Association that created textiles while working with regional artisans—along with Maddalena Forcella. Informed by a social anthropological approach, this textile design collective engaged with weavers and embroiderers from Mayan villages in San Juan Chamula and Tenejapa. Around 20 artisans once participated in the workshops that they held about natural dyes and traditional Mayan designs. The workshops assisted weavers on design, quality control, pedal loom, and dying techniques. Morales was also a friend of James Metcalf, and Ana and Jorge Pellicer—former CECATI 166 directors and cofounders at Santa Clara del Cobre.[6] My interview with Luis Morales centered on three points: his work at CECATI 133, his work with artisans as the cofounder of the association K'un, and his reflections on Mexican anthropological literature.

The fact that Luis was trained as an anthropologist with experience in social design and with artisans was of great interest to me. We discussed issues related to anthropological questions concerned with local contexts that might contribute to my theoretical framework for the project. I was also very interested in discussing works published by Mexican anthropologists, such as Beltran and Pozas, and which were only available in Spanish. During our conversations, we discussed work by Adolfo Best-Maugard and Dr. Moises Saenz (1888–1941), research developed in Chiapas by Calixta Guiteras Holmes and her publication

Perils of the Soul (1965), as well as widely published work by Dr. Robert Carter Wilson (1995).

Our discussion assisted me in gathering evidence and to develop a theoretical framework to address local circumstances. Morales went on to discuss the indigenist movement in detail, and we talked about the ides of Moisés Saenz, Bonfil Batalla, and Juan Villoro who all supported the Zapatista movement. After our interview which extended until dawn, Morales expressed his understanding of my research as a continuation of CECATI 166's initial spirit. He explained that he recognized similarities between this project and the K'ụn K'un project that he had started 30 years prior with Maddalena Forcella.

2014—November 1. Visit to Textile Museum in Oaxaca

I returned to Oaxaca, México, in October 27, 2014. I attended celebrations at the Textile Museum and at the local cemetery following local customs. The Day of the Death Celebration (Día de Muertos) is a Mexican holiday celebrated throughout México. Originally only celebrated in Central and Southern Mexico, it became a national holiday in the 1960s as part of the government's cultural homogenization policies. The holiday's roots can be traced to Mesoamerican cultures, and today it syncretizes a cosmological vision of life after death with the Christian celebration of the Day of all Deceased and All Saints' Day. Scholars trace the origins of his Mexican holiday to an ancient Aztec festival dedicated to the goddess Mictecacíhuatl.

On this holiday, it is customary that families honor their dead, pray, make offerings, and gather to help their deceased in their spiritual journey after death. In Mexican culture, this is viewed as a natural part of the human cycle. Día de los Muertos in México is a joyous day of reencounter with departed relatives and friends. Tradition calls for building private altars—called ofrendas—giving gifts, honoring the deceased using skulls, marigolds, food and beverages, visiting graves, and oftentimes holding celebrations in the Cemetery at night.

2014—November 6. Travel to San Cristóbal de las Casas, Chiapas

In my subsequent travels across México (see Figure 5.3), I learned about current political and cultural issues. Expressions through graffiti often revealed social protests' revindications of unsolved conflicts. One demand at the time included the recovery of a group of 43 missing students who were abducted on September 26, 2014 while they attended the Rural Teacher's College in Ayotzinapa, in the State of Guerrero. That region is one of the most damaged by drug trafficking In México. Investigations eventually resulted in the recovery of three corpses. The rest are still missing, yet recent independent investigations by a group of experts from the Inter-American Commission on Human Rights have found that the Mexican armed forces knew about and falsified evidence that could have helped locate the missing students. It has been found that the students were under heavy governmental surveillance due to their strong ties to

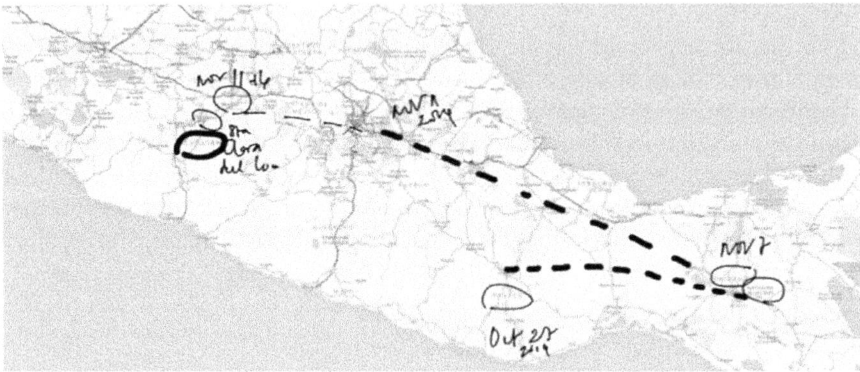

Figure 5.3 Drawing: Carmen Malvar. Map tracing travel conducted during this Ethnographic Research phase. CADA Foundation®, October and November 2014.

Note: The map/route shows the places where a visit with the intention of gathering data in preparation for the future workshops: Oaxaca City/San Cristóbal (Chiapas)/ Santa Clara del Cobre (Michoacán).

leftwing social movements. In spite of the official versions to close the case and that claim mass murder or incarceration by a drug cartel, it remains unclear who abducted the students and, presumably, killed them. In México, issues of politics, poverty, immigration, and drug and human trafficking have a strong impact on the social and individual well-being of its population.

2014—November 8. Visits to San Cristóbal and Michoacán. Interviews with Aurelia Zebadua, Luis Morales, and Ana Pellicer

During my second visit, Luis and I scheduled a visit to CECATI 133, as he was still serving as the Director (see Figure 5.4). During his tenure as Director, he tried to recover what had been achieved at Santa Clara del Cobre's CECATI 166. At the training center, he introduced me to a goldsmith and artisan named Aurelia Zebadua who had been a student under the apprenticeship of James Metcalf, and prior to the formation of CECATI 166. There, she learned jewelry techniques for casting with lost wax melting. Now she teaches methods of lost wax melting and we were able to discuss her craft and see samples of her work and techniques. This interview was recorded in video format and is partially edited online and by invitation-viewing on the Vimeo Platform: https://vimeo.com/489942190/b2e1afecc4.

Luis Morales would join me some days later in Michoacán for a joint interview I would hold with him and Ana Pellicer. This interview was recorded in video format and is partially edited online by invitation-viewing on the Vimeo Platform: https://vimeo.com/489872162/c084d5f4e8 (see video of *2014—November 15. Santa Clara del Cobre. Interview to Ana Pellicer and Luis Morales at Pellicer's house)*.

Figure 5.4 Interior spaces at CECATI 133 in San Cristóbal de las Casas, depicting metalsmithing artisans at work. San Cristóbal de las Casas, Chiapas. CADA Foundation®, November 2014.

2014—November 9. Travel to San Cristóbal de las Casas

I visited the El Camino de los Altos Association. The initiative was founded by French designers who worked with local weavers. There, I visited with some team members. Ana Paula and Maddalena had strong ties through their previous collaborative work there.[7]

2014—November 10. Travel to San Juan Chamula

During 2014, I also visited San Juan Chamula, Chiapas. It is a Tzotzil Mayan village in the highlands of Southern Mexico and outside San Cristóbal de las Casas. This community uses the system of *Sistemas Normativos Indígenas*, or a self-determined way of ruling.[8] Their approach to cosmological and oral tradition has been documented by many anthropologists, but these efforts have never been published in English.

Chamula inhabitants maintain their customs and dress in their traditional clothes. Women wear a beautiful local wool textile that they themselves spin, weave, and dye black. They felt and brushed it vigorously to create a fascinating textile which they use as skirt. Men wear a garment similar to a tunic made with the same fabric.

2014—November 11. Travel from San Cristóbal de las Casas, Chiapas, to Santa Clara del Cobre, Michoacán

My next travel by bus took me from San Cristóbal to Mexico City; a 15-hour road trip, overall. The journey would continue to Morelia, in the State of Michoacán, for another five hours. I would further drive to Santa Clara del Cobre, to meet Ana Pellicer at her residence. A long and worthwhile journey for our discussion.

2014—November 12. Santa Clara del Cobre. Meeting Ana Pellicer, Former Director of CECATI 166 Adolfo Best Maugard

I met and interviewed Ana Pellicer on November 12, 2014. The experience proved fascinating for the project. Ana contextualized past and present artisan situations, needs, state of affairs, and potentials. She spoke about the rich traditions and history of craftsmanship in México since pre-Columbian eras. Ana Pellicer also discussed the potential for how shared learning techniques may lead to the construction of collective knowledge and empowerment. She also shared a key bibliography with me that I could reference while developing my theoretical approach to the research. Such interview was recorded in video format and is partially edited online by invitation-viewing on the Vimeo Platform: https://vimeo.com/490532721/7609b3ec60.

My research gained a greater theoretical grounding from the interview with Ana Pellicer. She can be considered one of the ground breakers and main characters of the technical training efforts during the 1960s and 1970s. She is also an accomplished sculptor with pieces, such as La Máquina, at the Museo Nacional de Arte Moderno in México Distrito Federal. During the interview, Ana also spoke about the history of the book *Artisans of the Future* (Jorge Pellicer, 1995), as well as her work and life that she shared with James Metcalf and other like-minded intellectuals and anthropologists. Ana and James believed in the power of design and craft as a means to elevate Mexican artisans and expose their knowledge and art to the world. This also contributed to the artisans' generation of improved artifact quality which wielded more market value and allowed artisans to earn a higher standard of living through their craft.

Ana began her career in Mexico City. Later, she met and married James Metcalf and they moved to Santa Clara del Cobre where they partnered their research. Metcalf, the son of artists from New York, became an accomplished artist himself. He had trained at the Pennsylvania Academy of the Fine Arts in Philadelphia, the Central School of Arts and Crafts in London, and had studied ancient metallurgy in Mallorca. He had served in the Second World War and lost some of his fingers there. Between 1956 and 1965, Metcalf lived and worked in Paris. He arrived in México in 1966 after he was invited to mount an exhibition at the Palace of Beaux Arts in México Distrito Federal.

Metcalf would move again to Santa Clara del Cobre in 1967 with Ana Pellicer. He arrived there with an English edition of Theophilus' book, *The Various*

Arts: Medieval Treatise on Painting, Glasswork and Metalwork (1847), and a truck full of tools he had brought from his atelier in Paris. Metcalf fell in love with Santa Clara. He discovered crafting techniques dating back to pre-Colombian times in the rural community. Pellicer and Metcalf both found themselves immersed in the artisan community, working with them, experimenting and training. An interview from Dos Artistas en un Pueblo de Artesanos, by López Zuckermann, was partially edited online with her permission, and is available by invitation-viewing on the Vimeo Platform: https://vimeo.com/490238219/2d659ad825.

Both artists founded several schools in Santa Clara del Cobre. In 1973, they received the support of the Ministry of Popular Culture and opened La Casa del Artesano, where they offered training to Santa Clara coppersmiths beyond the traditional training they received as apprentices in local family workshops. The founders managed to provide a space for the community to perfect their traditions through experimentation, transformation, and the creation of daily utilitarian objects in silver and copper. La Casa del Artesano was closed in the late 1970s.

In 1976, Metcalf and Pellicer also began offering training in their home. Deep tensions, however, continued to exist within the artisans' community and between ancient and modern techniques, working styles, and community life in general. In the same year, Pellicer and Metcalf founded the Educational Action Center 67, which would later become CECATI 166 Adolfo Best Maugard under the auspices of México's Secretaría de Educación Pública (Secretary of Public Education). The organization introduced different techniques of metalsmithing and jewelry making at all levels of production. In doing so, the school incorporated both traditional and European forging methods, and introduced blacksmithing and casting in both lost wax and sand. They also introduced new machine tools, lathing, enamel work, stone cutting, and electroplating. All of those techniques opened new multifaceted horizons of artistic and commercial opportunity to Santa Clara artisans.

Both Ana and James also conducted their own work during their time in Santa Clara that contributed to the community's trust toward the couple. Ana, thinking of the town´s women, introduced jewelry craft lessons. While James became a well-regarded artist in México and his adoptive town, he and the community worked together to design the torch for the 1968 Olympic Games.

CECATI 166 became an unprecedented training center of its kind. Its aim was to provide training to artisans who wanted to explore their techniques of copper and silver metalsmithing, jewelry, and woodworking. Metcalf introduced new technology and techniques such as the thick edge and invented tools such as the turntable for leveling. Metcalf remained fascinated by local techniques that he had believed were lost in Europe some 5,000 years prior; yet they remained intact in the work and craft of Santa Clara's artisans. The center was also innovative in a broader community context since it was the sole training center to include a daycare center for the artisans' children. The school created strong communal ties among the artisans based on their shared knowledge, mutual respect, fairness toward each other, and a notion of collaboration.

In 2002, a Michoacán branch of México's Teachers' Union took over directorship of the school and displaced Metcalf and Pellicer. The move was highly politicized and its consequences spilled over into extreme community tensions and division between the copper artisans and the former directors of the school. The training center became closed by force due to intrigue and political reasons. Presumably, this was largely due to jealousy and envy toward the school's directors.

2014—November 13. Santa Clara del Cobre. Interview to Marina Fraga, Former Area Coordinator, CECATI 166 Adolfo Best Maugard (1995)

During my interview with Marina Fraga, she remembered directors James Metcalf and Ana Pellicer as wise and generous artists and facilitators who were dedicated to their crafts. She pointed out they had been mentors who were genuinely interested in training craftspeople in fine craftsmanship, teaching new techniques, mastering, perfecting, and validating ancient methods (see Figures 5.5 and 5.6). Marina remembers: "The artisan collective was happy … However, unnamed political schemers manipulated gullible community members to generate discontent among them." Fraga argues that the schemers spread rumors about rich people abusing the artisans; using them for their personal benefit as cheap labor. At one point, a riot was orchestrated by the political agents and destitution fell on then directors Metcalf and Pellicer.

Figure 5.5 Fine examples of wrought iron, zoomorphic door handles abstractions of a bull and an armadillo. Santa Clara del Cobre, Michoacán. Training samples in fine craftsmanship, teaching new techniques, mastering, perfecting, and validating ancient methods under Directors James Metcalf and Ana Pellicer. CADA Foundation®, November 2014.

Figure 5.6 Detailed craftsmanship in an edge with openwork and a pendant, examples of jewelry and the production of fine decorative objects. Santa Clara del Cobre, Michoacán. Under Directors James Metcalf and Ana Pellicer. CADA Foundation®, November 2014.

2014—November 14. Field Visit to Former CECATI 166 Adolfo Best Maugard. Interview with Carmen Pinoco and Director Miguel Ángel Alcantar García

I scheduled an anonymous visit to CECATI 166 at Santa Clara del Cobre after making a phone appointment without warning the current Director of my research interests. Ana Pellicer warned me of the situation at the school, and my aim was to reconfirm the state of affairs through a visit that should appear to be spontaneous. I did not tell officials at the training center about my research or my purpose for being there.

The image (see Figure 5.7) presented is a testimonial to the current condition and abandonment of CECATI 166 after it was closed by government officials and the forced resignation of James Metcalf and Ana Pellicer. My visit and Marina Fragas' interview are both edited online and are available by invitation-viewing on the Vimeo Platform: https://vimeo.com/489939510/f25c4cf831.

2014—November 15. Visit to Familia Hernández Cooper Workshop. Four-generation Artisan Family

While at Santa Clara, I also visited the Hernández family who are copper artisans. Ana Pellicer referred me to the family. They have been working with hammered copper over four generations, starting with Gilberto Hernández. The following three generations of blacksmiths and artisans work together and are now led by Gilberto's son Napoleón (65 years old), Saul (40), and José (12). Even at José's young age, he is already an apprentice to learn their craft of using an ancient pre-Hispanic process.

Ana knew the family well since she had once ceded machinery to them. These tools had belonged to her late husband and artist James Metcalf. The family's access to the tools allowed them to work at a new and different scale to

Figure 5.7 Still workshop at CECATI 166. Current condition and abandonment Santa Clara del Cobre, Michoacán. CADA Foundation®, November 2014.

produce artifacts. These objects are now exported and sold to an appreciating national and international audience.

The artisan family explained their techniques to me and showed me around their workshop. I recorded a video documenting their processes, which includes melting metal inside a pit in the ground filled with ash. As a gesture of gratitude, and in exchange for their hospitality during my visit, I edited and sent the video to them as I had promised.

2014—November 15. Santa Clara del Cobre. Interview to Ana Pellicer and Luis Morales at Pellicer's house

Given my previous interviews with Ana Pellicer and Luis Morales, I had gained a greater knowledge of social, political, and historical contexts in México as well as a greater theoretical grounding. Thus, I prepared a list of questions for another round of interviews with Ana Pellicer and Luis Morales; both of whom previously served as Directors of CECATI 166 and 133, respectively. These questions led to a deep conversation about Mexican tradition, history, and the development of technical and artisanal trainings in the country's political context. The interviews were filmed and edited. A video recording is included partially in this book's Meta-research. It was published by CADA Foundation® and is available under open password protection on Vimeo: https://vimeo.com/489872162/c084d5f4e8.

2014—November 16 and 17. Visits to Patzcuaro-Morelia and Mexico City

Luis Morales and I made a further visit to Patzcuaro in Michoacán. This town is well known for its crafts in metalworks, textiles, basketry, woodworking, lacquer, miniatures, toys, and stone carving. Eventually, the trip would be extended to México DIF so as to meet with fellow designers, gallerists, and museum curators.

2014—November 21. Oaxaca Close Field Work Final Stage.

After my travel and exploratory research in San Cristóbal, Michoacán, and Mexico City, I traveled back to Oaxaca to hold the first meeting with the designers who would represent a team of facilitators for incipient CADA Foundation.

The years 2013 and 2014 proved to be a very intense and fruitful period for my field work. I gained invaluable insight and connections through my travel experiences and discussions with people in the field. I could have never predicted that I would meet and interview so key figures involved in developing an approach to training artisans and who had experiences working closely with artisan communities. All of these figures held similar goals to promote the uniqueness and value of such craftsmanship to other artists and the general public. In these discussions, I benefitted from learning of the many sources about previous efforts that have never been translated into English or that have been poorly disseminated and credited in Spanish. My findings during this time generated many unforeseen synergies for me to gain perspective, knowledge, and understanding.

During this time, I also gathered a team of facilitators along the way who were already in contact with artisans and their communities for years. These connections helped lay the groundwork for developing oral workshops in rural artisan communities who expressed their own interests in experimenting with new materials and tools for greater access to new markets. The workshops, their locations, aims, achievements, and the relationships that developed between all participants will be described in the next section along with a discussion of how these factors affected design processes in this project.

Workshops and Facilitators' Meetings

The previous section presents my ethnographical field work that I conducted to meet various activists, anthropologists, artists, and educators who have extensive experience working with artisan communities in Oaxaca and Chiapas. I also interviewed gallerists and business owners who had specialist knowledge or interests in cultural artifacts, techniques, and craftsmanship in both the public and private sectors. This research involved extensive travel and maintaining a respectful, honest, and open perspective toward Mexican cultures and people.

This current section portrays detailed accounts of the workshops and meetings that participants conducted after my travel to gather information and form a stronger theoretical background for the project. I describe some of these instances as case studies where workshop development and outcomes are described in detail. Since I collected an extensive collection of recordings, videos, photos, and notes, I select some of what I felt were the most important experiences I had while conducting the research for this chapter. Some of these videos are available on open platforms such as Vimeo and have been cited throughout the document.

In the ensuing pages, I describe workshops and meetings in chronological order and which have been mentioned in Chapters 1 and 4. This section expands on my previous mentions of Phase I: Ethnographical Research and Community Selection, Phase II: Design Research, and Phase III: Prototyping and Artisanal Production for Market Exposure. Some workshops are fully documented as case studies.[9] Such Meta-research can surely assist other design practitioners in formulating and developing their own design methodologies adapted to circumstances and local cultural contexts in co-creation processes.

Phase I: Ethnographic Research. First Oral Workshops

Facilitators and I held a series of initial oral workshops with artisans in February 2014. These experiences served as ethnographic explorations in our first stage of mapping the territories in terms of geography and local context. I consider these workshops to be preliminary to the future design practice workshops.

Our aim in organizing these preliminary workshops was to create visual and narrative summaries that encompassed individual and community perspectives toward tacit knowledge, making things, and exchange. Simultaneously, they prompted stronger relationships between the facilitators and the community members so as to open a meaningful dialogue and mutual understanding of community goals for the future design workshops

In these oral workshops, women, children, and elderly from different communities expressed their oral traditions and mythologies through drawings which I summarized in a useful dossier. Images and legends containing figures born from symmetry of nature and abstraction of symbolic elements were collected in these oral narrative workshops. Through the images and legends collected, we shared a basis to exchange knowledge, and the facilitators and myself were allowed a cultural approximation for developing the future design workshops.

2014—February 8. Santa María de Atzompa, Oaxaca. Oral Workshop 01 (Half-Day). Facilitators Carmen Malvar, Omar Fernández, Maddalena Forcella, Fátima Díaz (Logistics)

Proximities of Design

Social: The town belongs to the Oaxaca Centre municipality, contains 21,000 inhabitants, and is located about 6 km from the center of the State. Zapotec and Spanish languages are spoken in the town, and no translator was needed. Artisans here are mostly dedicated to making pottery. The pottery is an important activity and products are exported to United States as well as local markets across México. All members of the family—including children—work on the production process as their main income source. An important strategy for their product marketing has been the artisans' introduction of a lead-free glazing process in the pottery production. The pottery is well known for its green color and glazing, and also for the artisans' use of the pastillage technique. Connections with community are made through Omar Fernández, who knew some of the artisans who expressed their interests in joining us for an oral narrative workshop to share their communal stories. Although the artisans expressed their desire to participate, we facilitators made sure to inform them that there would be no payment for their participation and expressed the academic purposes of the project.

Space: We hold the gathering inside an open patio. It is an official public municipal building. The team and the participating artisans (mostly women) take chairs and form a large open and inclusive circle. In the second part of the workshop, the tables are placed in the center of the circle and the artisans move chairs closer together (Figure 5.8). The use of space is constrained to within the area provided by the patio which was granted to us by local authorities.

Material: Plastic tables and chairs were provided to the workshop by the municipality. A coffee machine, coffee, milk cookies, sugar, water, cardboard cups, handmade paper from Etla paper factory, and color waxed crayons are used during the workshop. I requested permission to use a camera, audio, and a recording device to document the workshop. This included my explanation of the importance behind documenting the process both for the project's academic purposes and for their own record once we are able to return copies of the drawings to them.

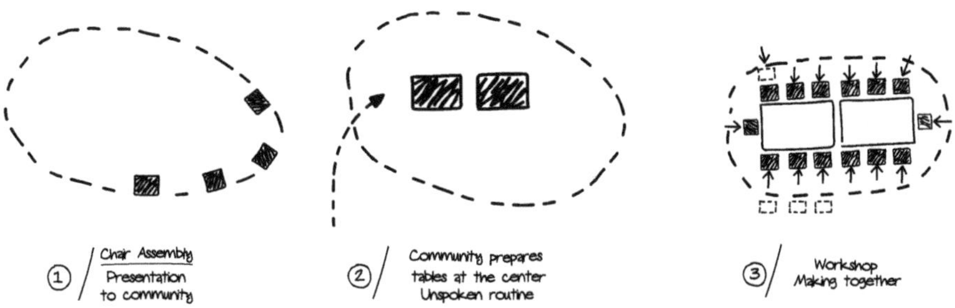

Figure 5.8 Diagram developed in field observation by Carmen Malvar. The circular dynamics of chairs and assembly with regard to space. Santa María de Atzompa, Oaxaca. CADA Foundation®, 2014.

Flow of Framed Events

Process: We facilitators approach the community through oral narrative relying upon the material discussed in Chapter 4 as a guide. We adapt the materials to the specific conditions of the workshop. The facilitators and I introduce the project to the participant in a simple and nearby way. First, we invite them to share stories or legends verbally. Then we facilitators share what we are doing as professionals and what we plan to do with upcoming workshops in the other communities who requested our support. We also ask if they would like to share the outcome drawings with all of the communities involved. The community expresses their interest in sharing and receiving other communities' drawings. Stories are told by various members of the community that have been orally passed from generation to generation. After a two-hour break to share coffee, the facilitators and I invite the artisans to draw their stories to transform their oral narrative into physical drawings by hand of their imagination (2D) with color crayons on paper.

Organizational Routine: First, facilitators and artisans sit in chairs. The sitting arrangement is led by the artisans and there are some children sat on their knees. Some notes on the verbal stories are taken by facilitators. The event lasts approximately two hours. In the second half of the workshop—nearly another two hours—artisans move tables to the center of the chair circle as an unspoken and organic organization routine. Facilitators distribute materials and coffee among participants to start second part of the workshop.

Dynamic: Around 30 artisans take part in the workshop to include older members of the community, middle-age adults (mostly women), and some children in a participatory manner. All participants engage in a discussion to create an atmosphere of trust and open exchange. As one participating woman begins telling community legends, the rest bring encouraging versions or add details. The environment becomes more relaxed and natural as dialogue opens and includes all participants. The team and participants from the community are enthusiastic. There is a total involvement of both children and adults. Participants share colored crayons. There is an informal closing message and a thank you message from facilitator team. We returned to share the final dossier with them.

2014—February 15. Santa María Yavesía, Oaxaca. Oral Workshop 02 (Half-Day). Facilitators Carmen Malvar, Omar Fernández, Maddalena Forcella, and Ludovica Morales

Proximities of Design

Social: Santa María Yavesía is a municipality of 472 inhabitants, and of which 27% speak an indigenous language. Zapotec and Spanish languages are spoken in the town. Fifty percent of the population possesses a primary education and only 15.95% has a secondary education. The municipality has an illiterate population of 10.48% and includes mostly older adults. This community has an extraordinary wealth of natural resources. The community is expecting us as they have expressed their interests in sharing their stories with us.

Space: The initial conversation with local authorities takes place in a large public space provided by the authorities. The community typically uses the space for music classes, workshops, and other activities.

Material: Workshop activities include the use of handmade paper from Oaxaca, ink, and brushes. The research team uses camera, audio, and a recording device to document the workshop after obtaining consent from the community through their governing body and with our promise to share the material back with them along with other communities' drawings.

Flow of Framed Events

Process: First, we arrive and speak with the authorities. The team and I introduce ourselves and the project with our academic reasons for visiting the community beyond responding to their request for support. After our introduction, we ask which stories the community would like to share in order to trigger a collective imagination. After a round of tales, we move on to the drawing portion of the workshop after the community expresses their interest in doing so. We close the workshop and promise to return with the results of the drawings compiled from other communities so that they can see how the other communities convey their oral stories and to compare or contrast aspects of each other's oral narratives.

Organizational Routine/Sharing Stories: First, the facilitators and I introduce ourselves, who we are, why we are here, and artisans tell their stories in a big circle of chairs. The chairs are prearranged by the authorities in preparation for our visit inside the community center. Later, and with the help of all the participants, circles are arranged differently around tables where adults and children transfer stories onto paper. We bring brushes and black ink with the paper that we bought at the San Agustín Etla factory. We work with the intention of knowing the community's stories coming from both adults and children.

Dynamic: About 40–50 members of this community take part to include grown men, women, and children. Community members are involved and enthusiastic toward participating as stories are told. A very positive and participatory response from the community. Conclusions are analyzed during the May 14 and October 28 facilitator meetings.

2014—February 22. San Pablo Tijaltepec, Oaxaca. Oral Workshop 03 (Half-Day). Facilitators Carmen Malvar, Omar Fernández, Fátima Díaz, Maddalena Forcella, Laura Quiroz, and Ludovica Morales

Proximities of Design

Social: San Pablo Tijaltepec is located within the Mixtecan region in the state of Oaxaca. With approximately 2,279 inhabitants, it is the most remote, the most marginal, and the least exposed to the market of the three communities I visited. As I have previously mentioned of the village, the artisans use embroidered textile with a pickup weave and gathered pleats process called pepenado. Local authorities rule by a self-governing system called "Sistemas Normativos Indígenas." This system is based in customary and traditional rule of law, where community leaders decide upon mandates, rotating obligations, and duties. Positions of power and decision-making are typically held by community men. Members of these communities are sometimes convened by either local authorities or government to attend meetings with foreign visitors. Their language is Mixtec. Translation is facilitated by Juanita Bautista and Laura Quiroz who maintain ties within the community.

Space: This workshop happens in the open air and within the public area of the community building offered by authorities, and is similar to workshops we held in the other communities.

Material: Workshop activities include the use of handmade paper from Etla, brushes, and black ink. Facilitators only use a camera because no recording is permitted by the community as they do not feel comfortable.

Flow of Framed Events

Process: We facilitators arrive and present ourselves to the authorities. We discuss our intention to respond to the community's interest in gaining market exposure for greater economic support through their crafts. This includes our discussion of creating a collaborative activity to trigger community's imagination as well as our interest to support the community's goals for market access. We start with an oral narrative activity including 30–40 women, and involves our discussion of legends and childhood stories which are then transferred onto paper with ink similar to previous workshops (Figure 5.9). Facilitator Laura Quiroz and artisan Juanita Bautista translate throughout the workshop.

Organizational Routine: Similar to the other workshops, the routine for gathering begins with the tendency of the participants to form a circle, although the existing space of the area is protected from the strong sun and does not allow a different set up for tables and chairs. We work around small

Figure 5.9 Woman holding her drawing at first oral workshop. San Pablo Tijaltepec, Oaxaca. CADA Foundation®, 2014.

tables provided by authorities. At one point, it becomes clear to us facilitators that there is a misunderstanding between ours and the authorities' message to the community about the workshops. This is because we connected with the community through the DIF (National System for the Integral Development of the Family) who offered the team and a car and driver for our community visits. We knew that the DIF depended directly on the local Oaxacan government, but we failed to realize that the artisans were typically obligated to participate in government-sponsored activities in exchange for payment. Under this patronage arrangement, artisans take their time with craft demonstrations in order to fulfill political agendas. Therefore, in this case, the women artisans felt that they were obligated to leave their work at home to join the workshops for alternative payment. Husbands then demanded money for their wives' time. In turn, we explained that we were not sent on behalf of the government. We explain that we are designers conducting research, and we are responding to the artisans' demand for support in their search for new ways to gain market exposure, but the situation already escalated. Juanita Bautista, one of our artisan contacts in the village, knows Lita Quiroz and supports us offering a more accurate translation and interpretation of the situation. The community, as a group, demand a high amount of money so they can return to their work places since some come from far in the highlands. Finally, after a tense situation, we reach an agreement where the van and the driver—originally lent to us by the DIF—makes several trips to bring some of the women closer to their homes together with their husbands.

Dynamic: This particular case generated a very uncomfortable situation. From the beginning, we facilitators could sense that some artisans were uncomfortable with the workshop dynamic although it took us time to figure out the problem. After the incident, we facilitators felt it would have been best to contact Juanita first rather than the local authorities. For future workshops, Juanita offered her house as a space for collaboration and offered to communicate with other women artisans in the community who were interested in participating.

Phase II: Design Research. General Reflective Meetings Part I

A series of preparatory and reflective meetings and discussions between project facilitators took place between May 2014 and February 2015 in Oaxaca Center after the oral workshops. These meetings happened before the actual design workshops scheduled for February and March 2015. The summaries shared here serve as an overview of those discussions and reflections. All main points of meetings are detailed. Each meeting duration was approximately two hours.

2014—May 14 and October 28. Oaxaca City. Facilitator Meetings. Facilitators Carmen Malvar (October 28), Rut Martin (May 14), Fátima, Kythzia, Diego, Omar, Maddalena, Ana Paula, and Laura

In these meetings, the facilitators and I classified and reorganized the information and reflected on our experiences we gathered during the previous workshops. We also revised the dossier of the drawings printed in black and white. Rut Martin took the time and spent his own resources to fly in from

Barcelona to support the dossier's development while I attended to personal matters in Galicia. This dossier document included the respective names and ages of all the participants who provided drawings. Each drawing was photographed, photocopied, labeled, and attached in the dossier. Copies of these documents were provided to each facilitator and the participating communities as we had promised.

These meetings also established the foundation for our development of the upcoming design workshops. Our discussions included methods for reflection and our views toward the role of "designer-facilitator" while working with the communities, failures, tensions, and potential future approaches.

In the meeting, we discussed avoiding the predetermination that participating communities faced any particular issue. This is typically a characteristic of the western vision that seeks to "fix" problems through imposing different beliefs without concern for local dynamics. Instead, we facilitators sought to approach the participating communities with caution and hear their goals and needs as they were self-recognized by the community members themselves. Since the artisans expressed their goals to increase global sales of their crafts, we facilitators discuss how my market strategy background may be useful during the workshops.

Importantly, we all focus on how design can be used to bring people together through more equitable processes. This involves recognizing the credibility of all opinions during the design and production processes. While we recognize our own needs as design practitioners, and seek a common approach based on skills and knowledge, we must also recognize the request of the artisans and their different skills and knowledge. We consider sharing the idea with artisans about approaches through new tangible materials in the artifacts crafting process.

The team and I take time during the meeting to reflect on our work so far and begin a process of self-criticism toward our inclinations to adopt positions of unconscious colonization since most of us come with aesthetic codes rooted in western influences. We agree that we must continue this process of self-criticism throughout the project and discuss our reflections during our meetings. We also plan to continue asking ourselves and artisans questions about the goals we aspire to achieve during the workshops in order to avoid this involuntary colonizing process. We are all in agreeance that the designer must leave the professional center and become a facilitator of a creative process within the community. This requires caution not to generate imposition, and it requires time for effective commitment (see Figure 5.10).

After our tense encounter with artisans in the previous phase, when they expected payment for participation in the workshop, we facilitators discuss strategies for more transparent invitations and communication for the future workshops. We feel that it is important to continue to adapt our roles as designers toward the artisans. We discuss the importance of maintaining our commitments to each community over time. This is because many designers, anthropologists, and historians have come and gone without reflecting on their impact and create false promises without any follow-up.

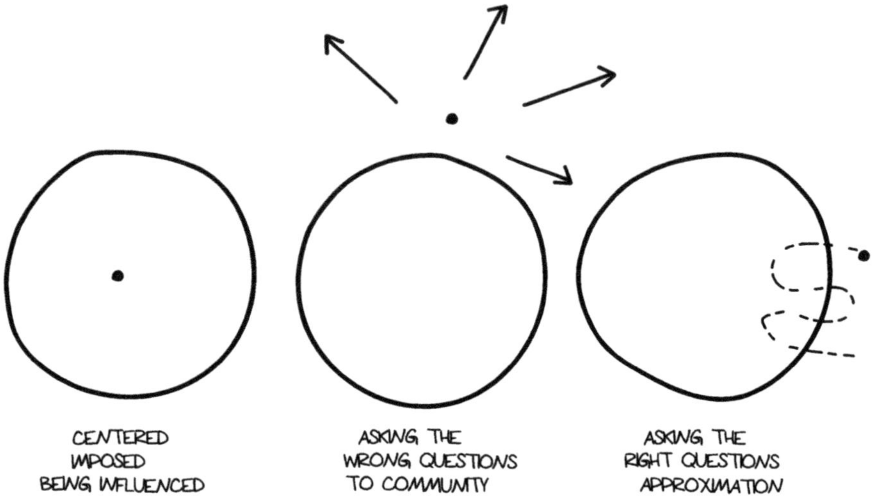

Figure 5.10 Diagram by Carmen Malvar showing off-center position in relation to communities in our approach. Left circle shows centered imposed position. Designer being of influence. Middle circle shows designer off-center asking the wrong questions to community. Right circle shows designer with an off-center position asking the right questions, creating a dialogue and approximation. Oaxaca Centre. CADA Foundation®, 2014.

In sum of this meeting, we more clearly defined the role and position of the facilitator for future workshops. We also proposed and clarified methods for collecting and analyzing collective community memories and histories. Facilitators and I also recognize the importance of materiality, time, and transformation in the processes of artisans' production. We share the idea that it is important not to set an end goal for the workshops. This is to say that we want to explore and learn with the artisans while also providing them our support. We also discuss the importance in informing artisans that the project is not government funded. We feel that it is important to be forthright and provide the community with our intentions to create and make together. Our primary objective is to support the artisans so that they may experiment and transform their craft techniques for eventually marketing their artifacts in a broader market.

2014—November 2 and 3. Oaxaca City. Facilitator Meetings. Founder of Colectivo 1050 Kythzia (November 2) and Facilitators Maddelena (November 3), Omar (November 3), and Carmen (November 2 and 3)

In the meeting on November 2, Kythzia Barrera and I revised images from toy selections and artifacts gathered from different cultures and across periods of history in order to share with artisans in upcoming workshops. We also revised the pottery typology from Santo Domingo Tonaltepec and the other samples from communities that work with clay in Oaxaca.

We discuss the proposal of Santo Domingo Tonaltepec to participate as a new community for workshops. This new proposed community will substitute for Santa María de Atzompa. We are aware that Santa María de Atzompa is already a commercial destination for tourism with existing market exposure. Santo Domingo Tonaltepec, on the other hand, has a marginal infrastructure. All main access roads are blocked from driveways and access to the community is difficult. The area is deforested by an old tradition of slash and burn. Unfortunately, the project does not have enough funding to work with both communities. Due to the logistical marginality and situation of Santo Domingo Tonaltepec, and the request of support from the community, we facilitators agree to develop workshops with the Santo Domingo Tonaltepec community. This village's artisans produce a pottery exactly as it was produced in the preconquest period, and the technique is in danger of disappearance.[10] I notify the artisans at the Santa María de Atzompa community of our change in plans to work with the new community, and Omar, who is close with the Santa María de Atzompa community, discusses the decision with them in more detail. The artisans understand and agree with our decision. They are also grateful to receive the dossier containing all drawings across participating communities from the first phase of the workshops. We schedule new workshop dates for Santo Domingo Tonaltepec, where some facilitators have connections through their work in Colectivo 1050. They propose a more intense and focused workshop process divided between three consecutive days. We discuss how to approach through the oral stories with a different materiality—clay—for this specific pottery community. This is because the community not only asked for strategies to gain market exposure, but also for support in the recovery of their crafts. Facilitators familiar with this community recognize the challenges of pottery scale. Santo Domingo Tonaltepec artisans create larger pieces with utilitarian purposes from the pre-Hispanic time, and the process has not been transformed since. Only two artisans remain who possess knowledge of the traditional process.

The next day, the facilitators and i revisit our latest notes and our next collaboration with the pottery community at Santo Domingo Tonaltepec. We also discuss relevant case study references focused in México and pertaining to social design and consider developing more information and research over existing case studies. We all find it difficult to gather material that could navigate us during the practice. We also schedule our next facilitator meeting between Maddelena and Laura to revise samples from San Pablo Tijaltepec brought by Laura to the Oaxaca Center.

2014—November 5. Oaxaca City. Facilitator Meeting. Facilitators Maddalena Forcella, Laura Quiroz, and Carmen Malvar

In this meeting, we facilitators revise some references of artifact images of toys from different cultures that I have begun to compile.[11] Also, the first samples have been provided from the community at San Pablo Tijaltepec, where we will be developing the next workshop on textile. We also discuss further references pertaining to the project.

Artisans have been participating in the previous workshops drawing their collective imaginary. The portfolio and Maddalena's previous work are reviewed during this meeting. We facilitators prepare for the coming workshops at San Pablo Tijaltepec. We analyze some of the samples from textile work provided by artisans Sebastiana García, Natividad García, Rosario Nicolás, and Rosalía Bautista. These artisans remain consistent participants throughout the project. We also begin to explore how we could proceed to move from 2D to 3D crafts as we discuss the importance of keeping in contact with the communities between workshops. We schedule tentative dates for the new workshops. Maddalena Forcella will be leading the next workshop with Ana Paula Fuentes, and Laura Quiroz will help with the translation from Mixtec language and provide a link with her own community.[12]

2014—November 20. Oaxaca City. Facilitator Meeting. Facilitators Carmen Malvar, Omar Fernández, Maddalena Forcella, Fátima Díaz, Kythzia Barrera, and Diego Mier y Terán

I return from a trip to Chiapas and Michoacán. The team and I have a conversation about the journey. We also discuss Santa Clara del Cobre as an example of community empowerment and review my interviews with Ana Pellicer. We consider her and Metcalf's method where the power of design and craft has elevated Mexican artisans and exposed their knowledge and art to the world while generating transformative artifacts, and some of them with greater market value. I share the case studies on social design in Michoacán and contrasted with documents and material that Luis Morales has provided me. Some documentation from Morales is based on training programs considered references for future studies on the impact of our upcoming workshops.

During the meeting, we review the dossier edited with the results from the oral workshops that Rut Martin had carefully prepared. We review again our methodology of approximation and how important it is to document the process and share our documented material with the communities. Once again, we discuss the importance of removing ourselves from the center as designers, and therefore we should not give our opinions nor our judgments. Our position as designers should be as far from the center as possible and should be collaborative. We reflect on the cultural and the social values found in our results from oral dossier drawings. The compilation of drawings in the dossier tells some stories from the villages and we consider them a good milestone to initiate an approximation in our design process: The iconography of nature is mentioned as a constant collective imaginary in all communities. Nature-related aspects (animal, landscape) are transcultural and timeless; they serve as possible elements of communal dialogue. There is no need to evaluate neither judge the stories, just to acknowledge them and to reflect on them. The coherence of those reflected values and how the artisan decides to honor them all are important factors to consider. How to extract those values and aesthetics and create parameters to unify diverse beliefs? How to create universal parameters that transcend both time (past, present, future) and culture? How do we keep this united sense of collectiveness? We revisit our time commitment to the communities and agree on a minimum of four years for a deep social engagement.

This served as an important meeting for our development of the upcoming design workshops. In sum, we acknowledged cultural and practical differences between ourselves, as facilitators, and the artisan communities. On the one hand, the facilitators are defined by the design process and aim to avoid giving meaning to sample studies. On the contrary, the artisan does not typically participate in that open process. As facilitators, we want to avoid imposing our ways of knowing and doing, and we feel the artisans' great knowledge and unique processes are points from which we should learn. It may be beneficial for all involved to start from an open question to create a more relaxed discussion. Our design process is framed by the stories of the people, their power, and hierarchies that they themselves establish and participate in.

We also continued our discussion of the team's concern toward time commitments, objectives, and communication with each community. We agree to take time to reflect and think critically about colonialism throughout the duration of the project. This involves facilitators communicating our intentions and the progress of the project transparently and clearly with all participating communities moving forward rather than rely on translating our intentions to the communities through local authority structures.

2015—February 4, 7, and 15. Santo Domingo Tonaltepec. Facilitator Meetings. Preparation for Design Workshops. Facilitators Carmen Malvar, Omar Fernández, Maddalena Forcella, Kythzia Barrera, Diego Mier y Terán, and Laura Quiroz

Under the suggestion of Kythzia and Diego from Colectivo 1050, and after a request from the community to develop and reinforce their crafts, Santo Domingo Tonaltepec is included as part of the project to replace Santa María de Atzompa. Omar informed the community at Santa María de Atzompa of the change. Workshops with the Santo Domingo Tonaltepec community were scheduled later and in relation to other design workshops at San Pablo Tijaltepec and Santa María Yavesía. The facilitators and I already had relevant information about all the communities after oral workshops, existing facilitator connections within the communities, and our reflective discussion meetings.

Based on the community's request to reevaluate their own pottery craft for continuity, we scheduled some additional meetings to revise our proximities of design for this specific community. These meetings took place in February 2015. In those meetings, we facilitators discuss how to approach this specific community through the oral tradition. A tentative structure is discussed based on the fact the community wants to reevaluate their own crafts because only two artisans, Margarita Cortés and Gregoria Cruz, continue practicing the traditional pottery technique. Debating ways of introduction to the specific workshops, we consider the community context and their tradition of working directly with material (3D) forms as an important factor. As we take notes on creating unifying activities through storytelling and narrative, we facilitators feel that it is important to think about what kind of stories the community might like to share with us. We discuss what kind of stories we could ask artisans so that we can invite them to participate.

Based on the team's collective years of experience as design practitioners, we reflect on past experiences and situations that may inform us on how to exchange skills with artisans so that community artifacts will make their way into the global market, as the artisans have expressed interest in doing. We ask the question "… are you interested?" to facilitate involvement in our process and gain consent for collaboration. Once more, we revisit the fundamental element of this project to develop trust over time. Colectivo 1050 has been already spending time with the community and so we hope to gain community trust through their members' support as facilitators. We facilitators discuss providing the artisans with references of artifacts that we have gathered from other markets and cultures as samples of popular toys and abstract designs in order to trigger dialogue. Thus, we consider how we will share our (conceptual) design tools (synthesis, analysis, references …) and how to create mutual exchange with artisans. Overall, in these meetings, we facilitators reflected on our compromises ahead, methods, and possible dynamics for the upcoming workshops at Santo Domingo Tonaltepec.

In the same meeting Maddalena Forcella expresses her doubts about our next workshop in San Pablo Tijaltepec working with textile. We have already completed the oral workshop and we are planning to do the first design workshops based on artisans' drawings. Maddalena also shares her concerns as the textile medium may challenge design transformation from 2D to 3D. We have to consider the complexity and the symbolism of their iconography about animals and landscape. We feel that the facilitators should share designer tools and work together with the artisans to transform their artifacts without losing their identity. This involves focusing on the scale and exploring new patterns. Therefore, we facilitators propose exploring a narrative beyond the existing context after considering the dossier from the oral workshops and the external artifact-toy references as tools to share with the artisans.

Phase II: Design Workshops Research—Proximities of Design

At this phase of the project, we facilitators introduced an experimental methodology to the artisans in a sequence of workshops—serving as case studies—based on a series of events or proximities of design by which we defined our courses of action. Facilitators and artisans produced design situations collaboratively to elicit transformations during organic workshop activities. The workshops aimed to draw upon artisan traditions so as to offer insight into craft and its transmission. Acts of act of making and unmaking proved a means to transform material and practices through manipulations of objects for practical uses that are also charged with cultural symbolisms.

The technique of each artisan was not determined by the existing materials or tools, but rather by their conceptual sense of community and their collective goals. During this design process, the artisans modify and reflect upon their individual and community's social preferences. This process of social design thus allows us to collectively and collaboratively reconceive of materials. Such complexity is inherent to the community structure and how it relates to artisan production; the process that we present causes a previously invisible structure to become visible within a community. Situations that occurred during the design workshops are described in the following case studies.

2015—February 11, 12, 13. Santo Domingo Tonaltepec. Design Workshop Part I. Case Study 01. Facilitators Carmen Malvar, Kythzia Barrera, Ludovica Morales, and Diego Mier y Terán

The facilitators held the first design workshop over the course of three consecutive days. In the Santo Domingo Tonaltepec community, the facilitators included women and children of several generations in the workshop since the entire community of women and children are eager to participate. We hear their stories and see their representations of those stories in their drawings and their clay creations. Details follow below and throughout this section. We began with an introduction and a narrative workshop on the first day. Then, we followed with a dialogue that is triggered by a compilation of images representing toy figure artifacts from different parts of the world. On the second day, the conversation moves toward the toy artifacts as universal symbols. On the third day, the attendees molded clay figures with their intention to fire and finalize their crafts. These consecutive days of workshops served as the starting point for the next phase. In that phase, creating tangible characters using specific crafts and relying on selected elements of the clay creations and drawings that represent the community's stories is a shared decision among all participants of the previous workshops. Materiality will articulate bonds between participating members of the workshop that are triggered through open dialogue and the exchange of stories. I share the workshop dynamics below based on the *proximities of design* approach, and in chronological order according to when each workshop occurred. Proximities of Design

Social: Santo Domingo Tonaltepec is one of the few communities where artisans produce pottery the same way as in the pre-Hispanic period. This is an example of cultural continuity. With around 300 inhabitants, and isolated from the broader Mexican society, Santo Domingo Tonaltepec still retains its idiosyncratic ceramic practices. In three of the community's villages—including Río Blanco Tonaltepec—the pigments come from the bark of certain trees which are rich in tannins. They use oak bark to produce a pigment that is then poured drop by drop over the clay pot or jar while it is still hot from the fire.

The liquid crackles as water evaporates on the hot clay surface, leaving only the red russet color of the tannins. The production of this pottery involves several dimensions of the artifact's cultural substance: the organization of the production, the way distribution is attempted and lastly, the use itself. As I have previously mentioned, only two artisans named Gregoria Cruz and Margarita Cortés worked with the clay using this technique at the time that we began collaborating with this community.

Space: The gathering happens outside and inside of public space offered by the municipality of Santo Domingo Tonaltepec. Both spaces are official public spaces. The team and the participating members of the community contribute in discussions.

Material: The materials present during the workshop include a coffee machine, coffee, milk cookies, sugar, water, cardboard cups, handmade paper from Etla paper factory, black ink, and brushes, camera, audio, recording device, cardboard bases to classify clay artifacts (A4 size), and the clay provided by artisans.

Flow of Framed Events: Process, Organizational Routine (OR), and Dynamic

Process, Day One: Facilitators introduce the first day with a game and by creating a human circle where everyone introduces their names. Groups then break away by ages. The team presents the project to the participants in a simple and nearby way. Children remain on the open soccer field and adults are invited to go inside the public building. Facilitators distribute the materials (paper, ink, and brushes) among all the children.

Facilitators ask the questions: "Tell me what is happiness for you … ", and "What kind of nature and animals inhabit in your community?" These questions prompt the children to then work with paper and then clay to create shapes and animals to represent their responses. Adults are invited to go inside the public building and we ask them the question: "what story you would you like to tell your children …?" This prompts the adults to work with 3D materials (clay) to represent some aspects of their responses. At midday, we break coffee before beginning the second part of the workshop. We gather all participants and adults to share their work to the group.

Process, Day Two: Facilitators share ideas about the day's activities and we all revise the work from the previous day together. I bring A4 cardboard to the children so that they can mount their artifacts on bases and mark their creations with their names. Adults continue with their stories and contextualize them with other elements in 3D. The children present their stories to the adults. Adults are inspired by the children's approach and their imaginative ideas. The children's presentation seems to trigger components in the adults' work. We facilitators continue to work with adults and children. We break the activities to share coffee. After approximately 30 minutes of break, facilitators introduce the images of global toy artifacts as a reference, and the conversation moves toward considering certain artifacts—such as toys—as universal cultural symbols.

We asked the question: "What brought your attention to this particular artifact?" and "Why do you choose it? Why do you like it?" Adults then work with a figure that they choose. Facilitators share the toy-artifact dossier with artisans and we all discuss specific images. This includes discussing the reasons why artisans like their selection, why it will be sold, its history, and how it can be adapted to clay.

Process, Day Three: The third day starts with breakfast and reflection on the previous workshop days among the facilitators. We facilitators are aware that we need to approximate to the community through their technique and their tradition. We are aware that the process of approximating the craft skill will be slow since only Gregoria and Margarita maintain the practice with traditional knowledge. We continue working with children and adults. Children explore outside in the soccer space with the facilitator Kythzia, who brings a glass ball as an artifact and mediator for dialogue to trigger curiosity. The activity was recorded in video format, and is partially edited online. It is available by invitation-viewing on the Vimeo Platform: https://vimeo.com/489931786/b49d285b76.

Facilitators ask the children about future travel: "Where would you like to travel?" The children recreate a trip into the future with clay by creating more figures and using the cardboard base with animals from the day before. Adults are invited to see the children's work and gain inspiration from their stories.

Organizational Routine (OR), Day One: Facilitators begin by creating a human circle and ask everyone to join in and play. We separate after the introduction. There are 35 kids that remain on the open space and sit in a big circle on the floor. As the children sit on the floor, some of the facilitators sit with them. A large group of adults organize inside as part of an unspoken routine. Artisans and facilitators move tables to the center of the space and create a circle with chairs. Meanwhile, the adults work directly with clay. At first, their artifacts are very big in scale. Adults present the pieces. They stand up while the others remain sitting close to the artisan while they present.

OR, Day Two: We invite the children to the open space on a soccer field. Adults are inside the official building. Adults' are inspired by the children's work and they begin to make their artifacts in smaller scale. Children return to their animal and stories. The cardboard bases allowed the children to frame their ideas in a tangible manner. The cardboard serves as a container where they place their artifacts and tell their story. We break to share coffee. After approximately 30 minutes, we facilitators introduce the references of toys and directed the conversation toward toys as universal symbols.

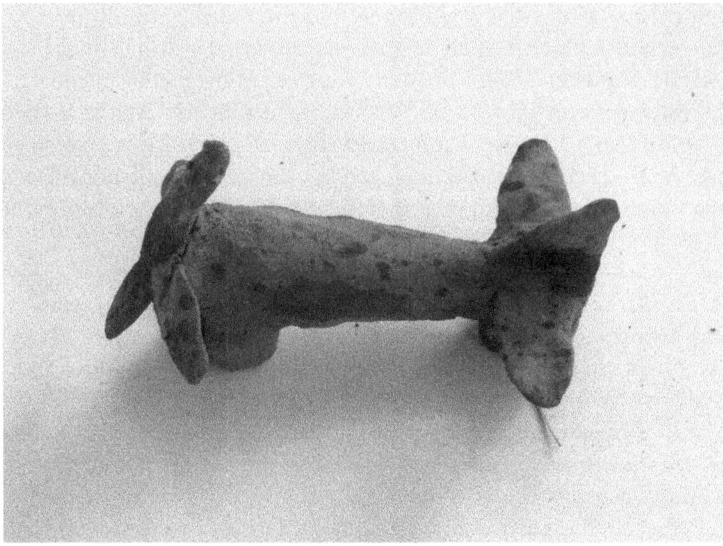

Figure 5.11 Children clay sample. Santo Domingo Tonaltepec. CADA Foundation®, 2015.

OR, Day Three: Children express more imagination in their stories and artifacts than the adults (see Figure 5.11). One child named Rocío is asked to explain her story to the adults and shows how the little piece of clay represents a huge imagination. The adults were invited to listen and use the children's imagination and scale to continue work on their own pieces. After the children's narrative, the adults were influenced to lower the scale of their pieces.

Dynamic, Day One: The team presents the project to the participants in a simple nearby way. All participants played with open attitude and a happy engagement. Some participants already knew facilitators Kythzia and Diego (Colectivo 1050). This made the atmosphere more relax and trustful. Children play outside with the clay and the adults work directly on clay as well. The entire pottery community translates their imaginative stories into 3D through clay. When we gather, all participants and adults present their work to the group.

Dynamic, Day Two: We ask participants to choose an image that they like from a series of images representing different artifact toys and figures, and that can also serve as universal symbols. All participants choose Mexican popular toys among avant-garde, or they chose images that are familiar within their own culture. Sharing coffee also contributes to creating a relaxed atmosphere. The facilitator team and participants from the community are enthusiastic.

Dynamic, Day Three: As the children introduce their stories, the rest bring encouraging versions or add details to their work. The environment is relaxed and natural. This facilitates dialogue among all participants. Artisans and facilitators agree to fire the clay figures so that we can proceed and match the time schedule that the other communities' workshops have reached. Facilitators schedule a meeting for review and conclusions on February 24, 2015.

2015—February 24. Santo Domingo Tonaltepec, Colectivo 1050. Facilitator Meetings. Facilitators Carmen Malvar, Kythzia Barrera, and Diego Mier y Terán

After the first design workshop at Santo Domingo Tonaltepec, facilitators meet to revise the results and to refine methods of approximating the community. We also discuss the community's request to provide pottery trainings for new and interested artisans. Colectivo 1050 plans to continue working with the community to bolster their commercial activities to provide greater access to markets.

In the meeting, we facilitators agree to hold an additional workshop in the Santo Domingo Tonaltepec community (February 26 and 27). We also revise and analyze artifacts developed in the previous workshop. We discuss revisions for workshop processes and dynamics as we plan future workshops. The children's participation in previous workshops was inspiring for the adults. We discuss the example of how one adult artisan, Gregoria Cruz, gained inspiration from one of the children's clay figurine.

In sum of this meeting, we reflected on the previous design workshop and analyzed results. We also talk about the possible approximation process for the next workshop at Santo Domingo Tonaltepec. This involves gathering and displaying all pieces created in the first series of workshops. We facilitators hope to develop a list of keywords to use in the next discussion in order to continue triggering open exchange and dynamics for experimentation. It is also important that we continue creating an equitable dialogue with the community to engage true and trustful relationships. We develop questions to ask the community about their hopes and expectations surrounding the workshops: "What kind of skills do you hope to develop during the workshop? what do you believe that you are gaining from the workshops?" Finally, during the meeting, we facilitators agree to emphasize that the artisans should not to be afraid to fail with new experimentations working with the clay materiality. This involves both facilitators and artisans simplifying forms and promoting the use of values in our work so that artisans may be more open to error without facing stigma.

2015—February 26 and 27. Santo Domingo Tonaltepec, Oaxaca. Design Workshop Part II. Case Study 01. Facilitators Carmen Malvar and Kythzia Barrera

The second series of design workshops at Santo Domingo Tonaltepec are held over the course of two consecutive days. Community members across several generations select work from the previous workshop as references to develop collective community characters and stories (Figure 5.12). We share some keywords ("happy," "sad," "drunk," "dead," "pregnant," etc.) with artisans to use in their descriptions and to trigger common stories. Then, we facilitators

Figure 5.12 Nazaria collecting the image from her work from previous workshop. Santo Domingo Tonaltepec. CADA Foundation®, February 2015.

Figure 5.13 Carmen and Gregoria Cruz reading through the artifact dossier during the trip back to Gregoria's home. Santo Domingo Tonaltepec. CADA Foundation®, February 2015.

collect the results (see Figure 5.13) as the artisans' schedule to fire their artifacts on March 9. This event marked the close of this phase for the Santo Domingo Tonaltepec workshops. Results are summarized in facilitator meetings between March 14 and May 3, 2015. The clay firing event is also recorded in video format, partially edited, and is available for invitation-viewing on the Vimeo Platform: https://vimeo.com/489911949/e8e8b43d79.

Figure 5.14 Sample from Sebastiana García. Textile sample commissioned for the design workshops at San Pablo Tijaltepec. CADA Foundation®, February 2015.

2015—March 2, 3, and 4. Santo Pablo Tijaltepec, Oaxaca. Design Workshop Part I. Case Study 02. Facilitators Carmen Malvar, Ludovica Morales, Maddalena Forcella, Laura Quiroz, and Ana Paula Fuentes

At Santo Pablo Tijaltepec, facilitators held the first design workshop over the course of three days. Women artisans in the community had already elaborated their artifact samples that were previously commissioned through Laura (see Figure 5.14). The community requested support for gaining market exposure, so it is important that we facilitators work to create a collaborative and transformative design process while maintaining a long-term engagement with the community. Drawings will serve as a basis for artisans to develop common characters and stories for their creations. The drawings will also serve as a starting point for creating tangible characters and elements using the community's embroidery technique to experiment with materiality and transform samples. Proximities of Design

Social: San Pablo Tijaltepec is located in the Mixtec region in the state of Oaxaca. The area contains approximately 2,279 inhabitants. It is very geographically isolated from other parts of México. Artisans are embroiderers and use a pickup weave and gathered pleat technique. The language most commonly spoken in the village is Mixtec.

Space: The gathering happens inside of a private space offered by one of the artisans named Juanita. Her house is in the village of San Pablo Tijaltepec. The exposed concrete walls are synonymous of well-being and prosperity. Some open holes for windows remain without any built-in frames. There is no running water and the toilet is outside of the house. There are a few chairs, a table, a fiber rug called "*tapete*," and a bed ready for our use in the working space.

Material: Among the materials used are a coffee machine, coffee, milk cookies, sugar, water, cardboard cups, the oral drawing dossier, and the printed and Xeroxed global artifact dossier. We facilitators also bring along some multi-colored thread, canvas fabric, wire, porexpan, plasticine, cork, felt, metal mesh, wooden sticks, and markers. The team brings a camera, audio and recording equipment to document the workshop. Artisans bring their embroidered textile samples previously commissioned.

Flow of Framed Events: Process, Organizational Routine (OR), and Dynamic

Process, Day One: On our way to the village, facilitators discuss some issues about privacy, our use of the camera, and the importance of explaining our presence clearly. We have been informed that this community has previously had unfavorable experiences with institutive initiatives that committed to provide equipment and training, but never followed up. After our previous misunderstanding with the community, artisan Juanita Bautista offered her house as a space for the design workshop. A total of seven women arrive on the first day. During the course of workshops, only Sebastiana García, Natividad García, Rosario Nicolás, and Rosalía Bautista remain committed to their interest in gaining market exposure and experiencing new techniques using different materials. Juanita explains that she could not commit due to her pregnancy and duties within the community. Juanita, Rosario, and Lita Quiroz translate Spanish to Mixtec for Sebastiana and Natividad. On the first day, we share our intention of experimenting together and exchanging knowledge so that the artisans can eventually gain market exposure and tools as per their desire. Our presence was a response to the community's own demand to market access. This time, our introduction is made with a thread and a skein, telling stories to each other as a way to gain proximity and share our plans to collaborate together. We facilitators also share the dossier containing the recompiled drawings from all previous oral workshops involving all communities. Artisans and facilitators share coffee and review and comment on the drawings; nature and animals are present in all drawings.

In the second part of the workshop, we all exchange materials and processes. We facilitators share some new materials brought from the city, and Juanita explains the pepenado technique to Maddelena and Ana Paula. We all experiment with each other's technique/material and make shapes in 3D. The materials are selected together with the artisans' own drawings and become the articulating bond in our collective dialogues. The new material and revisions of their own drawings and stories trigger the artisans' imagination while they share their knowledge to us facilitators through the art of embroidery. While we made the shapes together, we facilitators also inquired about their relationship with the artisanal embroidery technic. Facilitators asked questions such as: "If your embroidery could travel, where you would like it to go … Oaxaca, México, …?". As artisans responded, we facilitators also explained and talked about markets, products, families of products, means of payments, marketing, and the role of money in exchange.

We facilitators commissioned embroidered samples for the next day and suggesting that they could incorporate these samples into pieces for the second day of the workshop. We also clarified that we facilitators would pay for each sample piece at a set price, rather than for the time of attending the workshops so as to avoid the form of patronage which is often used by the government.

Process, Day Two: The team of facilitators meet in Tlaxiaco Market over breakfast to review the process for the day. I share some taxonomies with the team and suggest using the iconography of the community's imaginary as a method for triggering an exchange, and so that the artisans can work with figures that they recognize. We facilitators introduce the application of design tools, including analysis, experimentation, and references working through new materials and images from the global artifacts and toy figures collection. We hope to inspire the possibility of exchange with other artisan communities during the workshop. In the first exercise, facilitators propose that women tell a story for their children—similar to the activity at Santo Domingo Tonaltepec.

Each woman narrates a story that is eventually applied to a tangible character as we share ideas and design tools. After a short break and reflection, I fix images printed from the global artifacts collection on the concrete wall. Only an edited and diverse selection of images are presented. These include folk toys from different cultures, avant-garde puppets, and Asian decorative wood animals among others.

Women are invited to choose an image that they like from the wall. We review artifacts together and introduce the idea of the figures and toys as universal artifacts. We discuss the idea of a product belonging to a family of products and explain the objectives behind this strategy in the wider markets. Facilitators ask artisans the questions: "What brought your attention to this particular image?," "Why do you choose it?," "Why do you like it?" These questions were also formulated at Santo Domingo Tonaltepec. In a second round of informal questions, facilitators ask: "What catches your attention?," "How would you reproduce it?," "How would you apply your embroidery?" We talk about continuing next day.

Process, Day Three: Facilitators reflect while meeting over breakfast in Tlaxiaco Market. We share the idea to continue working with the same process as the previous day. We consider the idea of using the same trigger words as we used in Santo Domingo Tonaltepec workshop. In this particular community, we consider sharing the idea of working with a new play dough material to create facial expressions based on trigger words with the artisans because their embroidered artifacts are essentially in 2D form. As we experiment together, we all discuss whether or not the same process could be possible as artisans move to work with fabric. During the exercise, facilitators emphasize that the end artifact form is not important for this workshop in the hopes of inspiring experimentation as an important part of the process.

In the second part of the workshop, we mention the idea of pairing artifact figure forms: dog and cat, or man and dog. The group also discusses dates for the next commissioned samples. Laura will continue to visit the community in March and April for sample recollection and to maintain our collective dialogue over time. I share the workshop dynamic below based on our *proximities of design* approach.

OR, Day One: The artisans bring samples of their embroidery. Their small children always are with the women and stay throughout the workshop's time. Women bring a (plastic) bag where they keep samples and material for sewing and working on their embroideries. They are constantly returning to their embroidery to continue creating as conversations happen or breaks in the process occur. One artisan, Guadalupe, visits the workshop and realizes that the exercise is not a government activity where artisans are usually paid and offered free food for attending; she demands both from us. We explain once more that we were not sent by the government. Guadalupe does not return after the first day. Meanwhile, the rest of the group discuss drawings together and share stories one by one using a long thread in a game of turns to exchange voices and narratives through material.

In a reflection of new material that we brought from the city, we sit on the floor over palm carpet—called "petate"—that the artisans use to sleep and work (see Figure 5.15). We discuss the embroidery and one of the artisans named

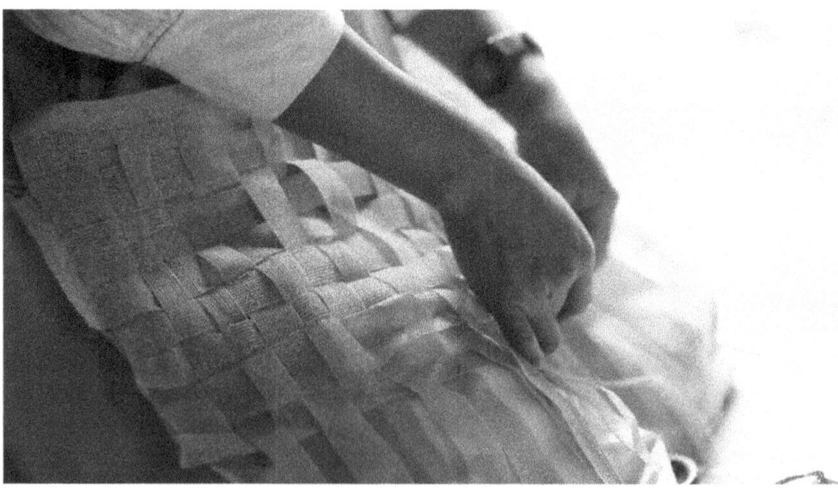

Figure 5.15 Experimenting with new materials. San Pablo Tijaltepec Design Workshop. CADA Foundation®, March 2015.

Juanita shares a demonstration. This demonstration was recorded in video format, is partially edited online, and is available by invitation-viewing on the Vimeo Platform: https://vimeo.com/489900016/def06d81cf.

When we ask questions about the city, many of the women explain that they have never left the mountains where they live. Most did not know what a museum store looks like. When we commissioned embroidered samples for next day, we knew that the offer would be of their interest since they constantly tried to sell their pieces to us. We clarified to the artisans that we would pay for each sample piece and not the time that they spent at the workshop. We wanted to be sure to establish a different sort of relationship with the artisans than the government and other initiatives which typically foster patronage dynamics.

OR, Day Two: Tlaxiaco is a city closer to Oaxaca Centre and a market destination for communities on the mountain. I booked rooms for all facilitators in a small hotel. It is the closest place to San Pablo Tijaltepec where we can stay and rest. The Tlaxiaco Market is a place for gathering, eating, and discussing in preparation for our day ahead. Facilitators review the process for the day, analyze samples, and identify respective cosmology and beliefs of the work over breakfast at the market in the morning.

Sharing taxonomies with the team allows the facilitators to enumerate trigger elements for the coming practice. We gather in Juanita's house as we did the previous day. Artisans Sebastiana, Natividad, Rosalía, and Rosario arrive at the same time as usual for a smooth routine. Facilitator Ludovica Morales prepares the collected images of selected universal toys and places them on the wall while the rest of facilitators sit and talk with artisans exploring their characters from first exercise.

OR, Day Three: During our explorations with tridimensional forms through creating facial expressions with the play dough, artisans also incorporate yarn and cloth to bring out their characters' facial compositions. The same process is applied with the animal forms. We revise the work together and facilitators suggest synthesizing the volumes and expressions of new forms in the same way. These forms include the artisans' interpretations of mountains, which are used in their frets, and which they later include on their sleeves.

When we commissioned embroidered samples, we once more clarified to the artisans that we would pay for each sample piece. Since the collaboration was proposed as an exchange of knowledge and ideas to reinforce their exposure to the market without losing their identity, we hoped to avoid recreating the patronage relationship that the artisans have experienced with the government and other initiatives. Commissioned samples are made to reinforce the idea of continuity and commitment through time as a collaborative design project.

Dynamic, Day One: When we facilitators introduced ourselves, the women demonstrate a great curiosity about us as family member. They wanted to know if we have a "man" with us, if we "have children," and if so, "how many children," as well as "how old we were." Artisans initially show shyness and a lack of trust. As we continue sharing the growing thread, the women start to smile and play along. When we asked, "if your embroidery would travel ...?" they show surprise and uncertainty. The lack of shared social/cultural reference points was a big difference between facilitators and artisans. The artisans showed a great uncertainty over the concept of sale, the market, and money. Their identification of their own drawings in a book, as they considered the printed dossier, and their curiosity shown over the other communities' work was valuable and important to note.

Dynamic, Day Two: The inviting questions are asked as a way to establish conversation and dialogue. The women artisans are hesitant to answer questions and proceed. As a cultural barrier, language differences and translations make it difficult for the artisans to ask facilitators questions. Artisans quietly observe and then choose an artifact image reference posted on the wall that represents universal symbols from various cultures. It is a quiet process, and the artisans show both curiosity and uncertainty.

Dynamic, Day Three: The play-dough facial expression exercise takes the artisans by playful surprise, and they react happily (see Figure 5.16). They are more engaged with this a 3D-making activity. Each day the complicity grows although facilitators still have a long way to go in order to gain the community's trust. The artisans react very positively to the idea of us facilitators coming back after the third day. The artisans also show engagement with the commissioned fabrics that Lita will bring to Oaxaca City in the following weeks. We facilitators

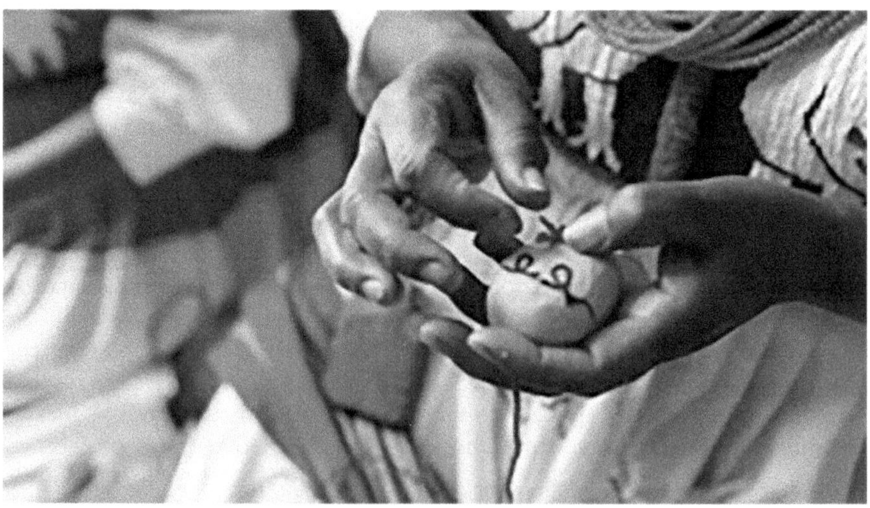

Figure 5.16 Working with 3D/play dough on face expressions. San Pablo Tijaltepec Design Workshop. CADA Foundation®, March 2015.

schedule a meeting for review and conclusions on March 14, 2015 with the team in Oaxaca City.

Phase II: Design Research. General Reflective Meetings Part II

Between March 14 and May 3, 2015, facilitators met in Oaxaca Centre after the first series of design workshops to review and reflect on the process and results in preparation for a second round of workshops. Each meeting lasted for approximately two hours and were often held over food in both public and private spaces. These reflective meetings and discussions contributed to our development of revised processes for the next phase of the project. I share summaries of these meetings below and provide the main points. I also documented these meetings using audio and video recording, and photography.

2015—March 14. Oaxaca City. Facilitator Meeting. Facilitators Carmen, Fátima, Diego, Omar, Maddalena, Ana Paula, and Laura

In this initial reflective meeting, facilitators analyze the work of the previous workshops. We study methods of approximation for the future workshops based on our reflections. Facilitators also analyze and discuss ways to formulate new questions for the artisans. Our revisions of the questions are based on ways in which facilitators may avoid imposing their beliefs or practices in workshop exercises and, in turn, formulate appropriate questions to facilitate more open discussions. This is necessary because the project aims to trigger design situations without imposing foreign beliefs and practices about design in the local communities. Some cases, when the artisans ask very specific questions, have proved to be more difficult for the facilitators to elaborate on a narrative without imposition. Our development of a process that avoids such forms of imposition will take time and collaboration with each community to so as to understand their sense of time, trust, and values. Thus, we facilitators agree that our questions ought to be broad and flexible. We consider shortening our questions if necessary and provide examples instead. Facilitators also revise the circle diagram representing the designer's role and position (see Figure 5.8). Facilitators discuss our options to develop a process of interaction without generating an imposed change in each community.

Facilitators agree that each community had gained more trust in the facilitators and the practice by the third day of the initial workshops. We feel that this is most likely due to our collective discussions over the course of the three days in each community. The artisans, such as Rosario Nicolás at Santo Domingo Tijaltepec, started sharing their personal issues with facilitators as time went on and discussions continued. When we facilitators were at Rosario's home, she elaborated on her journey to learn embroidery and felt that she was not proficient even after eight years of practice. From situations such as these, the facilitators gained greater understandings of time values and the dialogue of trust.

In the meeting, we facilitators also talked about the consequences associated with paying for the artisan's samples. Maddalena mentions how, as in the community of San Pablo Tijaltepec, there is pressure and a need for selling garments for money payment. The artisans begin to elaborate on further samples than are commissioned. Sebastiana, Rosario, Natividad, and Rosalía request more money with each visit as they provide new samples for payment without consulting the facilitators. Therefore, facilitators agree that we must be clearer and take more time to explain the purpose of producing samples. We also must discuss differences in purpose between the concept, process, and finished materials with the artisans. Rosario supported our discussions since she is fluent in Mixtec and living in the community, but she has also lived in the city and has knowledge of other market systems. Moving forward, we are all together in total agreement that suggestions for artifact products must come from the artisans' ideas.

Facilitators comment on types of markets and their processes in our group meetings. We believe that it will be important to discuss these concepts together with artisans and collectively think about why they wish to have greater market exposure in order to sell their artifact products. Such a discussion might shed light on how market exposure could ameliorate the social consequences of limited economic resources faced across generations of community members. Facilitators revisit the case of Xochistlahuaca,[13] where Maddalena and Ana Paula have collaborated in the past, and discuss how the community worked specifically with colors and design over the course of two years. During that time, women became aware of their possibilities and organized as a group. We mention how this sample is a successful case as artisans gained more confidence with time.

2015—March 17 and May 3. Oaxaca City. Facilitator Meeting. Facilitators Carmen, Fátima, Diego, Omar, Maddalena, Ana Paula, Kythzia, and Laura

As facilitator debates continued, we laid out the entire collection of resultant samples to discuss and review aspects of scale, issues of size and weight, and measure the relevancy of combining pieces produced across communities. We found that the initial plan to add material elements from across different communities would increase costs and complicate logistics. We facilitators revised our notes again based on our discussions with artisans over scale, production, and the possibility of producing puppet style toys for increased market exposure. We talked about the production processes ahead: how many pieces we could commission, their sizes and scales, and the potential additional costs. We also discussed our doubts as we faced a limited budget. As artisans continued to work independently to produce artifact samples, we committed to meet regularly as a facilitator team. We facilitators continued to emphasize the importance of creating unique, but replicable artifacts to the artisans for gaining market exposure.

It remained important to us facilitates that the project's development over time was a result of the artisan's—and their respective communities'—goals and needs. These goals also informed the research goals, which remained dedicated to providing support for the artisans in order to challenge global production pressures and bolster local resilience for each community. Thus, we facilitators agreed to refuse the DIF's car loan and driver as part of the project's resources. We are clear that neither the DIF, nor any other politically aligned organization, should be involved with the project moving forward. Such an affiliation would confuse the goals of the artisan communities and this project. This decision had a severe impact on the project's already low budget, and created tensions with DIF until a new governing body took office.

Between all this time, we facilitators continued collecting commissioned samples directly from the artisan communities. These samples continued to align with the parameters that were established in our previous exchanges between artisans and facilitators. We all agreed to revisit in June after the next series of workshops.

Phase III: Design Research. Prototyping and Production

Until this point in time, the design workshops catalyzed dialogue, generated questions, and facilitated new relationships. As I have previously mentioned, the interaction among participants bolstered dialogues about techniques, materials, and the remaking of artifacts. This democratic activity gives voice to local people. Macro aspirations of the community are reflected upon, discussed openly, and expanded further. The artisans' ability to experiment and transform is highlighted in the workshops through their use of the oral storytelling, their applications in drawings, and their creation of artifacts. All of this also served to trigger new forms of interaction between artisans, and between artisans and facilitators, which culminated in our collective exploration of new skills related to the creation of new textile and pottery artifacts to come.

2015—May and June. San Pablo Tijaltepec and Santo Domingo Tonaltepec. Phase III. Facilitator Meetings and Workshops

After the previous meeting, we facilitators moved forward with another phase of the project to focus on prototyping and production for commercial marketing and sales. On May 3, 2015, we facilitators meet over a video call to discuss our next steps. Maddalena, Ana Paula, Omar, Kythzia, Laura, and I are present for this discussion.

In our meeting, we facilitators once again review the topics of scale and market access. Also, we discuss concerns about the monetary and time expenses involved with the next phase. We realize that the project will need further funding in order to proceed. In this phase of the project, we facilitators start a new stage of developing prototypes with the artisans. The workshops involve sharing tools that allow the artisans to integrate their production of artifacts into a globalized market following the interests of the community. Based on previous workshops, the artisans recognize their goals and define the social design practice as a possible tool to building resilient identities.

In this phase, our goal as design facilitators is to begin outlining samples and prototypes. The resulting design process developed across the previous workshops had already established our collective dialogue about the use of different techniques and their application to different materials. Our aim to accomplish this collaborative work remains focused around fostering the strength and resiliency of the artisan communities without altering the character of their traditions and collective identities. We also commission the artisans to produce bags for packaging the artifacts similar to those produced at El Camino lo Los Altos.

As experienced design practitioners, we facilitators have a firm grasp on the concepts of sales, cost calculation, revenues, and how a serial product could impact success. The prototypes that artisans produce during this phase offer an answer based on the aesthetics, form, interaction, usefulness, and emotional content of the community; they serve as models to establish and develop the preparation techniques for a collection of marketable products.

Due to the project's constrained budget, we facilitators were unable to link communities in a combined effort to market a collection of products where each community would elaborate on their particular artifact. But, we facilitators could commission new samples at Santo Domingo Tonaltepec and San Pablo Tijaltepec. Therefore, Diego returns to Santo Domingo Tonaltepec with an assistant to pick up previously commissioned samples and to hold another series of workshop on June 2 and 3, 2015. Diego and his assistant Fernanda follow up on the production in previous workshops, and they talk again with the artisans Gregoria, Demetria, Nazaria, Antonia, and Margarita about pricing strategies and the market expectations for consistency within that frame. After showing the artisans photographs of their creations from the first burning and contained in the dossier, Diego and his assistant ask: "how do you feel to see these works now?" In response, the artisans are critical of their objects' scale and voice their concerns that "larger objects will be more expensive to move."

Diego introduces the subject of the market and shares some images of utilitarian, decorative ceramic objects (see Figure 5.17). He asks the artisans: "what is the difference that you can tell between those finished objects in the markets and the product that you have developed so far?" The artisans reflect together upon seeing their objects in the dossier and look for the utility in their object while improving the decorative details. They also talk again about the concept of a product family. At the end of the two-day production workshops, the participants then select their own artifacts for firing in the kiln.

On June 17, 2015, facilitators meet in my rented space at Calle Hidalgo in Oaxaca City. Lita Quiroz brings samples from San Pablo Tijaltepec for our collective evaluation. Maddalena, Lita, Ana Pula, and I are astonished by the vast quantity of items; from finger puppets to small bags to handheld toys. We review this array of artifacts and share our idea for Natividad, Sebastiana, Rosario, and Rosalía to focus on fauna since animals are represented in all embroideries created for garments and samples. We are deeply moved by the impressive sense of creativity in the artisans' transformed production processes.

Figure 5.17 Artisan Nazaria and facilitator Diego, revising utilitarian and decorative commercial pottery at workshops in Santo Domingo Tonaltepec. CADA Foundation®, June 2015.

We facilitators discuss how to address packaging for sale again as we prepare to travel to the community over the next two days for a fourth visit and a new workshop series.

During June 18 and 19, Maddalena, Ana Paula, Laura, and I conduct another workshop. Rosalía offers her house and furniture for our visit (see Figure 5.18). In the workshop, we facilitators review samples of dolls and animals with Natividad, Rosalía, Rosario, and Sebastiana. Together, we discuss ideas about how to use embroidery for future packaging and future applications in their samples. In order to support the artisans' future productions, Maddalena introduces the concept of templates to continue reproducing the animals that begin to have 3D qualities. The women still sew the animal pieces by hand, although Rosario possessed a broken sewing machine. We eventually recruit a seamstress from Oaxaca Center, named Rufina Ruíz, who is happy to become a facilitator for this portion of the project. We facilitators meet with Rufina several times in Oaxaca Center to review Rosalía's, Natividad's, Sebastiana's, and Rosario's 3D animal fabrications. We consider the ways in which we can collaborate to transform production for greater efficiency without sacrificing the strong senses of identity contained in each hand-crafted piece. We invite Rufina to San Pablo Tijaltepec over the following months to train the artisan women with an extra sewing machine as Rufina fixes Rosario's broken machine. Rufina repairs the machine and shares the process with Natividad, Sebastiana, Rosalía,

Figure 5.18 The house of Rosalía Bautista prepared to hold workshops. Oaxaca Centre. CADA Foundation®, 2015.

and Rosario. Rufina also shows the artisans how to assemble 3D patterns and how to insert filters to create volume (see Figures 5.19, 5.20, and 5.21). During the workshops, the artisans produce many samples as they begin to experiment with the applications of their own embroideries at different stages of their 3D figures (see Figure 5.22).

On June 21, we facilitators meet once again to discuss outcomes and review the process. We arrived to another inflection point for the workshop methodology, and discuss the issue of scale and packaging for bringing the artifacts to market. We are aware of the big leap in development that the artisans at San Pablo Tijaltepec have made from narrative to abstraction and using new methods and materials. The artisans have shifted from creating embroidered animals in 2D to creating in 3D. These animals serve as important symbols in the community, and the artisans have transformed their characters from the embroidery of the blouses. The artisans have addressed scale, finishing qualities, and now packaging. We share the idea with artisans not to condition the design process to commercial criteria and work under the editing parameters of the existing samples made so far.

We analyze all pottery, textile, and some wood samples and discuss how to proceed. Diego and Kythzia express conflicts with their other project agendas.

Figure 5.19 Rufina Ruíz and the exchange of tools and techniques with artisans at the workshop. San Pablo Tijaltepec. CADA Foundation®, 2016.

Figure 5.20 Natividad and Sebastiana García taking patterns out of shapes. San Pablo Tijaltepec. CADA Foundation®, March 2016.

Figure 5.21 Natividad and Sebastiana García reusing their old sewing machine arranged by Rufina Ruíz. San Pablo Tijaltepec. CADA Foundation®, March 2016.

The question as to whether or not the communities will manufacture an artifact component and a shared package causes another debate over logistical issues. The project will need greater funding in order for us to proceed with a new phase.

The biggest achievement that we can clearly identify is that artisans have appropriated their prototypes as extensions of themselves, projecting from their bodies. We facilitators believe that we have succeeded in refraining from imposing a colonial western perspective over the beliefs and processes of the artisan communities. Although we have exchanged tools with the artisans, they

Figure 5.22 Cuts and shapes developed at workshops with women of San Pablo Tijaltepec. CADA Foundation®, June 2015.

have made it possible for the artisans to decide on methods to transform their design and achieve their own goals.

Phase III: Design Research. Artisanal Production and Initiated Market Exposure

Between *October 2015 and March 2017*, facilitators held workshops at Santo Domingo Tonaltepec and San Pablo Tijaltepec. The artisans in Santo Domingo Tonaltepec and San Pablo Tijaltepec had clear objectives for production and goals for a concise market focus. On October 17 and 25, 2015, we facilitators met with two seamstresses in Oaxaca to address packaging issues. Also, we all were looking to define a strategy to implement patterns for cutting shapes in order to replicate them without scale variation. Our first meeting was at El Casa,[14] a cultural center sponsored by former Mexican artist Francisco Toledo. The center has an atelier of production well known for their afelpado[15] technic. Artisans had prepared some packaging samples that facilitators later disregard due to cost and our diminished budget.

In our second meeting on *October 25, 2015*, Rufina met with facilitators at Maddalena's studio. She has collaborated with Maddalena and was handy with a sewing machine. Rufina also had experience with weaving and embroidering, as well as a deep knowledge of fabric elaboration with a loom. Rufina brought ideas and suggestions for solving finishing issues and for improving the figures'

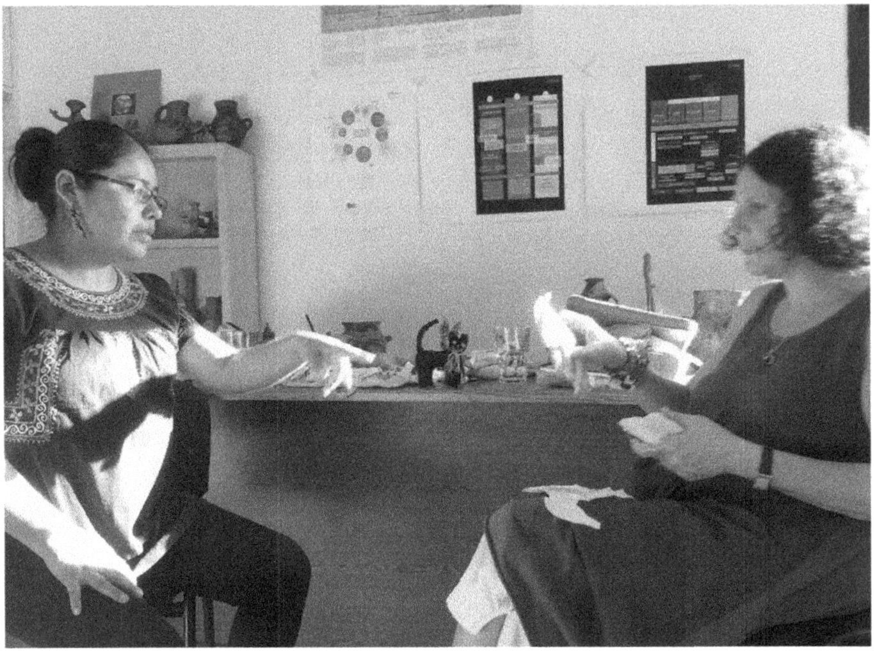

Figure 5.23 Laura Quiroz and Maddalena Forcella reviewing new samples commissioned and brought by Laura from San Pablo Tijaltepec. HUB, Oaxaca Centre. CADA Foundation®, 2015.

volume. Later, she shares these ideas with Natividad, Sebastiana, Rosario, and Rosalía in San Pablo Tijaltepec.

Later in *October 2015*, Laura, Maddalena, and I meet. We celebrate the following meetings at the HUB office space I rented in Oaxaca (see Figure 5.23). The space is often used by multidisciplinary artists and start-up companies in the center of town. I have collected a large number of books, samples, and material there over the course of the project. Laura brings more animal figures from her last visit to San Pablo Tijaltepec. We are excited with the artisans' development of the figures. Laura Quiroz agrees to return to the artisans and commission more paid samples. We have another meeting with Rufina, where she and Maddalena discuss possibilities for finishing processes to share with the artisans.

On *November 18 and 19, 2015*, we facilitators hold a two-day workshop at Santo Domingo Tonaltepec to elaborate on finishes, bring new ceramist tools, soften edges, and provide packaging measurements to revise scale (see Figure 5.24). We gather to work around a table full of samples. Gregoria, Margarita, Demetria, Nazaria, and Antonia are commissioned to produce four more sets of animal couples. I fund the pieces and their firing as well (see Figure 5.25). All material

Figure 5.24 Artisan Demetria Cruz working on the production of pieces. Santo Domingo Tonaltepec Workshops. CADA Foundation®, November 2015.

Figure 5.25 Firing. A group gather and communal activity. Samples burned and in process for burning. Santo Domingo Tonaltepec Workshops. CADA Foundation®, November 2015.

Figure 5.26 Result of some early samples already burned and finished. Some of the production was lost by breakage of the parts. HUB Oaxaca Centre. CADA Foundation®, November 2015.

results are brought to the HUB office in Oaxaca on November 26 and 28, 2015, for the facilitator team's comments (see Figure 5.26). Before going back to the communities, we revise samples to determine what finish had the most potential for success in the markets (see Figures 5.27 and 5.28).

On *March 4 and 5 2016*, we facilitators hold a two-day workshop at San Pablo Tijaltepec to elaborate on finishes in preparation for the upcoming production of samples. Rosario, Sebastiana, Natividad, and Rosalía have formed and consolidated a collective called Ñaa Ñanga meaning "women of the toys" in Mixtec (see Figures 5.29 and 5.30).

On *March 7 and 8, 2016*, we facilitators also held a two-day workshop at Santo Domingo Tonaltepec. We hosted a CADA workshop shared with Colectivo 1050. We revised the fired samples that I had commissioned. Colectivo 1050 invited an artisan from a nearby region to share her experiences with pottery and finishes based on using different types of clay. She presented her work and offered a demonstration. We facilitators felt that it was important to share techniques to preserve and empower local cultures with other crafts workers in the region. Such event was recorded in video format and is partially edited online and is accessible by invitation-viewing on the Vimeo Platform: https://vimeo.com/489922999/75bb96dd7e.

Figure 5.27 Diego reviewing fired pieces commissioned. HUB Oaxaca Centre. CADA Foundation®, November 2015.

Figure 5.28 Interior of a section of the temporary office of CADA. HUB Oaxaca Centre. CADA Foundation®, November 2015.

Figure 5.29 Working together. Exchange of tools and techniques with artisans. San Pablo Tijaltepec Workshop. CADA Foundation®, March 2016.

Figure 5.30 Some samples of resulting prototypes from San Pablo Tijaltepec with the use of sewing machine for finishing. San Pablo Tijaltepec. CADA Foundation®, March 2016.

In *October 2016*, Ñaa Ñanga was invited to the Second Mesoamerican Textile Congress (TEXTIM II) and Expo-Sale at Oaxaca de Juárez by the Textile Museum of Oaxaca, which provides a marketing opportunity for the women of San Pablo Tijaltepec. Participating at the TEXTIM II was a mesmerizing experience for the women. They ask me for an advance loan for their production to assure their access to materials. I agree to commission their work so that they can fund their production of 100 textile toys (25 animals from each artisan). They also ask me to provide a loan for a new sewing machine which I did with a contribution from Antonio Lorente. Antonio (Tucho) is a family friend from Spain who lent me the money to allow the effort to continue, as I had run out of my own resources during the project.

Prior to Ñaa Ñanga's invitation to participate in the textile congress, Natividad and Sebastiana had never left their village. The opportunity inspired both fear and excitement within the community, but also gave way to their sense of empowerment to challenge existing family structures and social roles. To attend the congress, the artisan women left their husbands in charge of their families, homes, and cattle for some days. The experience prompted the artisan women to challenge their initial prejudices and gain broader exposure through participation at the TEXTIM II and Expo-Sale at Oaxaca de Juárez. As I have mentioned, the event was organized by the Oaxacan Textile Museum, and many textile collectors asked to commission garments from Natividad, Sebastiana, Rosalía, and Rosario. The garments that the artisans brought and showed garnered a large interest from the collectors due to the artisans' embroidery techniques. Visitors at the fair showed their fascination for the artisans' garments and toy collections.

In *October 2016*, while the TEXTIM II and Expo-Sale at Oaxaca de Juárez was held at Museo Textil de Oaxaca, artisans at Santo Domingo Tonaltepec continued collaborating with Colectivo 1050 as my resources were already very limited. I found that it would be impossible for me to continue working on the project without any funding support. Colectivo 1050, however, had already initiated new workshops with the community. I remained engaged in different initiatives through the CADA Foundation to raise funding and awareness for the continuity of the project. The following year, in February 2017, I traveled back to México to attend an exhibit at the Hotel Galería Casa Comtesse. Artisans' sales were successful, and the Hotel office paid for a small selection of items to provide in their gift shop. By then, the Museum Store in Oaxaca had sold out of the artisans' pieces that were marketed at the Textile Congress.

On *March 17 and 18, 2017*, I developed a pilot training workshop together with Ana Paula and Laura at the Textile Museum in Oaxaca. We presented the project's study of Santo Domingo Tonaltepec and San Pablo Tijaltepec. Around ten local designers and cultural liaisons attended, and they shared mutual interest in the project since they all had conducted work with artisan communities. In the following days, the team and I met again to organize our information. By then, Laura had produced a video to start a crowdfunding initiative for the Ñaa Ñanga collective, and raised nearly $3,000 USD.

We returned to San Pablo Tijaltepec in March 2017 in response to Ñaa Ñanga's request for workshop support on sales, productive chain processes, and commerce. We also received an invitation for a celebrative meal of Pozole[16] at Sebastiana's house (see Figure 5.31). For the first time since we began the workshops, men joined our meeting. The men started to realize the women's earning potential generated through their crafts. We also began to realize how the project might have an impact on domestic power balances. We explained how to evaluate and project sales and how to charge commission on sales held in their premises. The artisans asked me to bring Rufina again to support artisans as they improve their blouses' proportions, lengths, and sizes with the sewing machine.

Figure 5.31 Artisan Sebastiana García. Member of the collective Ñaa Ñanga. CADA Foundation®, March 2017.

Eventually, we held one more workshop at San Pablo Tijaltepec on *July 27, 2017*. Designer and research facilitator Rut Martin traveled from Barcelona under her own expense to join Ana Paula and Laura. They worked on documenting the outcomes, revising finishes and sizing, and creating inventory. Laura organized another sale for the artisans and a further workshop at the Oaxaca Centre for October 2017. Rosalía, one of the younger artisans, also began a technique demonstration at one of the city's sales. The artisans also succeeded in getting samples of their work into permanent exhibitions at the Textile Museum of Oaxaca.

In 2018, I returned once again to Oaxaca. I tried to recover and make an inventory of clay pieces that I had commissioned. I traveled alone from Oaxaca to San Pablo Tijaltepec to meet Laura at Tlaxiaco and visit Rosalía who had dedicated a space to work and sew at home. We went over the results of the animal toys and also brought some samples of textiles from El Camino de los Altos (see Figure 5.32). We met at Rosalía's home one last time to share Pozole. Everyone—facilitators and community men and women—shared the food under trees near Rosalía's house. It was the first time that I felt as if I were not a stranger to the community.

Figure 5.32 Women in the 2018 workshop observing a textile from El Camino de los Altos at the new construction of Rosalía's house. New women who have been invited to participate arrived. San Pablo Tijaltepec. CADA Foundation®, June 2018.

Social Design: One Way to Approach Local Culture

How did we approach local culture through social design? In short, this was a slow and arduous process. The indigenous artisan communities involved in this project have connections rooted to place, limited economic resources, and a deep relationship between daily routine and cultural traditions and values. Taken all together, this makes for deeply bonded communities. Therefore, our approach required time, measured expectations, deep reflection, trust, and an openness to unexpected outcomes.

We facilitators and the artisans developed methods that were defined through careful interpretation and an evolving workshop process. Through discussion with other design practitioners, and by adhering to a reflexive facilitator role, the team and I developed a self-orienting process of discovery and collaboration during "*el hacer*"—the making. It was the artisans' goal to reach new markets, and the facilitators' goal to provide the artisans an environment where they were able to consider their craft, their collective goals, and make their own transformations based on tools that we design practitioners could offer. Rather than impose our knowledge of making, we offered the communities a service to expand the potentials of their goods and craft techniques based on their realization of their own needs and goals.

Together, we facilitators and artisans considered physical properties such as texture, geometry (size and location in space), and materials (weight, flexibility, and quality). We also considered dynamic properties as material artifacts could engage with all senses. The workshops were dependent on those early approximation case studies and the detailed reflections on previous meetings that I described in the early part of this chapter. Our discussions with artisans and our team reflections during those stages all helped participants make tangible sense of new worldviews. We were also able to chart a path forward based on material exploration. We encouraged the artisans to take control and experiment without expectations of the outcomes. Along the way, we all experienced technical challenges since these were new materials and forms for us all. More importantly, though, materiality was not only a technical property but an element to trigger ideas, exploration, and dialogue among all. The process was helpful for all participants, and it also exposed our own problems and insecurities. Many of the artisans felt frustrated and shared problems that they encountered with the material. We responded with advice and suggestions. Facilitators also felt insecure and tensions rise. But, through hearing each other's suggestions during our collaborative processes, the artisans were able to elaborate a stunning range of samples and gain market exposure without sacrificing their local identities. For us designers, we understood the artisans' creations as artifacts that embodied a process of cultural transmission through "*el hacer*"—the making.

Notes

1 Encyclopedia of the Municipalities and Delegations of México. Corresponding section to the State of Oaxaca: www.inafed.gob.mx/work/enciclopedia/EMM20oaxaca/index.html

2 Locality catalogs of Oaxaca. Microregion Unit. www.microrregiones.gob.mx/catloc/Default.aspx?buscar=1&tipo=nombre&campo=mun&valor=&varent=20

3 Silk production in San Pedro Cajonos http://viernestradicional.impacto.org.mx/blog/la-produccion-de-seda-en-el-san-pedro-cajonos/

4 A parasitic insect of the Nopales.

5 See more information about Ambar Past in Chapter 4.

6 During my documentation of Oaxaca and the production workshops in October 2013, I found the book "Los artesanos del porvenir" (The Artisans of the future) (Pellicer, 1995). This brought a new perspective to the research became a key reference in the project. Pellicer's book is based on a case study at Santa Clara del Cobre and discusses a project planned and conducted through CECATI No. 166 by James Metcalf and Ana Pellicer. In the work, they assert and reclaim artisan production as opposed to industrial processes not only for its collective value but as the motor of local reality. The authors posit that artisan production is the only valid precondition for true industry based on implementations of basic crafts and supporting crafts. Pellicer's book revisits the work of his close collaborator James Metcalf and describes one of the best documented case studies known about artifact production in México (and more precisely in Santa Clara del Cobre). Discussing the book with Maddalena Forcella would spark an introduction to Luis Morales because he is a close friend of Jorge and Ana Pellicer and James Metcalf. Morales tried to implement the same approach to educational structure and echoed the work previously done at CECATI 166 which is presented in Pellicer's book. Metcalf's, Morales', and the Pellicer's work contributes to a relevant sample of artisan production and the reinforcement of communities through their design of arts and crafts.

7 See more information about Maddalena and *El Camino de los Altos Association* in Chapter 2, footnote 36.

8 See more information about *Sistemas Normativos Indígenas* in Chapter 4, footnote 33.

9 Initial workshops' participants were in large numbers, and I was unable to take note of all of their names for this book. As the workshops progressed, fewer artisan remained involved and I introduce their names as part of the whole group of collaborators who remain active and in relationship to the CADA Foundation or through facilitators until the point that this book is published.

10 The craft is only practiced today by two artisans (Margarita Cortés and Gregoria Cruz), and knowledge has not been transmitted to new generations because many younger people have left the village to study or find work. Potters have difficulty commercializing their products since the community is located far from main communications, roads, and markets. Few are dedicated to selling some of the pots in local markets, and there is no agriculture beyond the minimum subsistence.

11 See more information on this reference book in Chapter 4.

12 There are many variations of Mixtec language.

13 La Flor de Xochistlahuaca is a cooperative of Amuzga women from southern Guerrero. They make textiles with traditional designs and transmit the traditional art of the backstrap loom among the new generations of the municipality of Xochistlahuaca: www.facebook.com/LaFlorDeXochistlahuaca

14 Located in the town of San Agustín Etla, and just 15 kilometers north of the City of Oaxaca, CASA defines itself as a plural forum that strengthens reflection and analysis. The CASA, founded by Francisco Toledo, is focused on artistic experimentation, exhibitions, and workshops. https://oaxaca-mio.com/atrac_turisticos/vistahermosa.htm

15 Afelpado is the process that transforms woven wool into a dense fabric with the fibers closely attached to each other. Wet flocking uses heat, agitation, and moisture to shrink and compact the wool fibers. www.handweavers.com/the-wool-process

16 Pozole is a traditional and very popular recipe or dish of Mexican cuisine or gastronomy. The recipe consists of a broth or soup made of corn kernels (originally a variety of peanuts) to which ingredients such as meat, pork, and chicken are added. Vegetables and spices are also commonly added which vary by region. https://recetapozole.com/

References

Guiteras Holmes, C. 1965. Los peligros del alma: visión del mundo de un tzotzil. México, DF: Fondo de Cultura Económica.

Pellicer, J. 1995. Artesanos del Porvenir. México, DF: Secretaría de Educación Pública, Universidad Autónoma Metropolitana.

Closing Chapter

Exploring Continuity, Measuring Impact, and the Legacy and Future of Social Design Practice

From my initial travels in Oaxaca, the project spanned nearly five years. Collaborators at all levels of this work set forth a great effort to put social design into practice. We traveled long distances, and often across difficult paths, to provide our services and exchange craft knowledge with local Oaxacan artisans. The economic costs of our coordinating logistics—i.e. the purchase of materials for workshops and travel expenses—was a key factor contributing to the project's longevity. It took us facilitators a long time to complete all the phases of this work since it was necessary to pause until we could produce a large enough budget for each next phase. During those pauses, we planned and revised our methods to develop more effective workshops that focused on the needs and goals of the artisan communities.

Participation in the project was formed in somewhat of a natural way; the facilitators, craftsmen, and collaborators were all those who were interested in the project's aims and who valued the opportunity for cultural enrichment and shared knowledge. Many began as complete strangers who learned to trust each other and learn from one another. The formation of these relationships was vital for the project's success. There were also some of those who wanted to join the team but could not establish commitments with the rest or who did not share the values of those already involved. This was also important for our discussions as a team, and helped to provide clarity during our meetings where facilitators and artisans planned and coordinated future efforts. Ultimately, the fact that our project's deadlines were extended due to logistical issues allowed us facilitators to gain greater connections with the communities and create a project with more meaningful long-term commitments. The project's pacing proved to align more-so with a comfortable timeline for the communities to feel confident in their contributions and fostered greater trust between all participants.

The research process was a learning experience for all involved. The research team's initial expectations about the project evolved and became more defined as time went on. We focused less on quantity of sites, the number of workshops, the amount of time, and resources spent. We realized that, from a qualitative point of view, the project was stronger and became more involved with the community. Human relationships were tightly woven through the power of "making" together over the years. Bonds between all participants grew stronger and stronger

DOI: 10.4324/9781003360803-8

as we shared food, stories, and time with each other. While our expectations shifted, our goals to implement a social design practice to understand local identities and build community resilience in the global age remained the same. In this chapter, I summarize the project's aims and discuss its broader impact.

Revisiting the Project Goals

During the course of the research, I could define three main objectives. These have to do with approaching communities, exploring the possibilities for using design to understand culture, and considering how design can facilitate equitable connections between local cultures and the global network. I conducted the ethnographic work with reference to the existing literature on such topics to gain context and perspective.

I was finding the appropriate pathways for approaching local cultures through a social design framework. The use of craft as a social design practice provided a system for exchange between designers and artisans. The designer's role in this context of "making together" was to provide a service. Through open discussion and a collaborative workshop environment, each community was able to share their worldviews, recognize their goals, and record their oral histories. The facilitators offered tools and new techniques that the artisans could use to transform their craft and realize their goals. The workshops were organized in order to manage the appropriate social, material, and spatial "proximities of design" in a democratic environment of trust and commitment that allowed facilitators to approach local cultures and avoid assuming an imposing role. Rather than claim expert knowledge as designers often do in the Western tradition, the workshops operated to build environments for mutual knowledge production and sharing between artisans and designers. The aim was to use design as a tool for community empowerment and combat colonization.

The ethnographic research that I conducted in Phase I of the project served to be instrumental to our ability to understand local communities' cultures as design practitioners. Through our approximations of local customs, values, and worldviews, the research team was able to form greater cultural understandings and respectfully approach the communities. Interaction between different cultures is a key element in this research since our aim was not to impose a specific methodology or worldview during the workshops, but rather to exchange different knowledge forms shared between the artisans and the team. Facilitators and collaborators that took part in the process were designers primarily born in México or who had lived there for a long time. These team members had also previously worked with indigenous communities and had different professional backgrounds and international trainings. In some cases, they joined the workshops with connections to the artisans of nearby communities to share their techniques, their way of working, and the materials they used. This allowed the group to work from a cultural approach that was familiar and opened avenues for open dialogue. In our team meetings, we shared our questions and resolved issues democratically. These characteristics of the team were key to getting into practice our established guidelines that avoid a Western colonizing approach.

We facilitators also found it important to act as reflexive practitioners while we approached the communities. We continually learned through observation to establish the proximities of design in each workshop and adapted our framing of the events as we reflected on the workshops. Through our reflections, we designers aimed to act with reference to the specific contexts of each community and aimed to avoid relying on some pre-determined criterion for developing the workshops. The position of the facilitator during the workshops was not to impose, but to promote dialogue and the participation of all. Through our use of the artifact and its making process, the workshops provided a space in which everyone felt free to express their opinions, to ask questions and make proposals, to give their perspective as equals, and to promote the exchange of knowledge and tools.

I believe that one of the crucial aspects of the team's research approach was that we did not establish an end goal for the research in which event we would then call an end to the work. Rather, our objective was the development of the process itself. The team focused on developing the workshops so as to facilitate discussions between facilitators and artisans to share ideas, knowledge, emotions, and memories over the course of those long years. While timeframe may seem over-extended from the transitory culture of capitalism in which we are all immersed, I believe that the longevity of this project was key to establishing a firm commitment between the entire team of facilitators and the communities. Our prolonged interactions gave credibility to the project and security to the participants.

I recognize this project as exploring how practices can help us all to better know and understand local identity in the age of globalism. Looking back in light of completing the workshops, I believe that I have been able to share how design practice can help understand local identities. In the workshops, facilitators were exposed to the oral traditions of each community, and artisans began to ask questions about their own cultural references after some discussion and drawing activities. The cultural artifact served as a stepping stone for discussion between the team and the artisans, and also as point for reflection for both. Artisans and facilitators established dialogues and exchanged skills, and through using traditional and new techniques and materials, the artisans were able to incorporate traditional culture in a series of prototypes. It was important that the team worked to establish the appropriate context of equality and correspondence during the workshops as that environment allowed for creating situations for mutual learning and an exchange of values and knowledge. In one example, artisans learned new techniques of pattern-making and sewing that allowed them to turn their drawings representing their oral traditions into prototypes of 3D figures. After reflecting on their oral stories and through learning how to operate a sewing machine, the artisans decorated their artifacts with the graphic and figurative iconographics that they traditionally embroidered on their blouses.

The role of the facilitator in these workshops was intended to be non-intrusive, and their aim was to facilitate the appropriate space for community decisions to be made through dialogue and equality. Different "proximities of design" situations were created during the workshops regarding the space, the people involved and the materials they used. These changed not only across the different workshops but also during the course of each workshop. Parameters such as trust, exchange or dialogue also shifted and transformed as time moved forward. This variation in the "proximities of design" influenced the team's decision-making during the design process as consequence of a better knowledge and understanding toward the local communities that we worked with.

As a result of the designers' reflective practitioner role, they were able to learn about the local culture in each different workshop through their observations and approximations of the artisans' participation. These proximities evolved to promote trust and dialogue among all participants—design practitioners and artisans alike. For example, the space where the workshops were held changed from initially being within a public community space provided by the local authority to eventually being held in one of the artisan's homes. First, we facilitators brought food to the public space, and eventually the artisans shared traditional meals with us facilitators which they prepared. In this context, the artisans' trust toward the team evolved, and new fields of dialogue opened to bring on new commitments allowing design decisions to develop further. An exploration of these changing situations, or "proximities of design," helped us all—the team and the artisans—to better understand local and global identities through place, people, and materials during our exchange of techniques and materials.

The increased trust between artisans and the team is reflected in our dialogue over the course of the workshops as artisans began to share their perspectives about the design process along with their own personal memories, emotions, and experiences. These more personal discussions opened new paths for interaction between all participants and provided us all with a wider perspective about each other, our concerns and expectations. Some of these discussions pushed us design practitioners to reflect on the impact of our research—not only among the individual experiences of the participants, but also in the social or economic organization of the community—and helped us to better understand the potential transformations to their local identities. For example, Sebastiana, one of the artisans from San Pablo Tijaltepec, shared with the team that she would like to buy a car if she could make enough money from selling her products.

One challenge that the team encountered was to bridge the design practitioners' vision to create more equitable opportunities in each community with the artisans´ vision to maintain specific roles within their community. In particular, in the case study at San Pablo Tijjaltepec, design practitioners observed the development of exclusive circles within the community that limited access to new ideas, training or materials the team provided in the workshops. Some community members sought to reserve the benefits of the workshops to a smaller artisan group and used their experiences in the workshops as competitive advantage against other artisans working in the same community.

This project also explored the ways in which design can support local communities in their relations to global markets. Turok and Novelo link expressions of cultural knowledge embedded in "*el hacer*" (the making) with the social and economic spheres that Malinowski (1957) had addressed years before. Their work generates a common thread among forms of artisan production, productive composition of the community, and relationships between local and national economies (Turok, 1988). This thread is particularly important regarding power imbalances that are evident in many social design practices which open the door to cultural imperialism. To avoid this, researchers and design practitioners should more closely consider the varying contexts at each local level and be guided by the human relationships found in each context. Trust and mutual interests between the intervening practitioner and the artisan can be found over time and through respectful cultural approximations on either side that go beyond commercial opportunism.

This social design practice approach based on trust, equality, and open dialogue can act as a guideline for workshops that aim to benefit local communities in the context of globalism. The specific cultural heritage of each community, however, must serve as a guideline within the workshop because it is ultimately up to the artisan to create an artifact which belongs to them and represents their community's worldview and interests. In this way, the artisans' artifact creations can foster a sense of community empowerment and belonging rather than a feeling of exploitation for Western consumers and industry.

As to ways in which design can support local communities in the global market, I believe that the broader implementation of a system for exchange that recognizes local economies and identities is necessary. Such a transformation in how powerful Western practitioners, researchers, and institutions approach local communities is possible. In this project, the team's design practice resulted in the artisans' design and materialization of artifacts that maintain the traditional techniques and images of their community while avoiding an intrusive cultural imposition on the part of the practitioners. Acting as facilitators, we design practitioners were able to establish commitments and create the appropriate situations to mutually exchange tools and knowledge. In the end, the communities were able to make decisions about what they might change or maintain in their design and production processes according to their own interests. Ultimately, the artisans were able to build resiliency toward the global markets through their incorporation of different techniques within their traditional practices.

In their discussions with the facilitator team, the artisans made proposals and suggestions for how they could transform their artifacts and connect to the global market. For example, the facilitators responded to the artisans' requests for information about how to produce their artifacts for commercialization and how to use new technologies such as social media to promote and manage their sales. Since the close of the project's workshops, Rosalía, Rosario, Natividad, and Sebastiana have managed their own sale of the goods and have transformed their family routines in the process, as men take on a larger role in childcare and women increase their family's income.

Within the prevailing non-Western discourse, social design is framed as a possible bolster for resilient and transformed identities within a context of globalization and the associated conflict between "modernism" and heritage. Within the context of this project, I believe that social design represents a possible tool for building resilient identities. The broader discourse about the role of design in developing countries has come a long way since the Ahmedabad Declaration of 1979. However, while the precedent for discourse centering around identity and social practice has been formed, a detailed critique of past design practices is still necessary to shape the future for social design. Technological approaches hold much potential in that they can help to create "new spaces" for small communities to generate big alliances and create a network of connections while also allowing cultural heritage to play an active role in community building.

Continuity, Impact, Legacy, and the Future for Social Design Practice

In retrospect of this project, I believe that it is very important for future work to establish long-term collaborations in order to strengthen the participation among all participants and to allow time for meaningful reflection. Although this research had to end eventually, I was not able to provide any formal closure to any of the communities with which the team and I worked. It was not possible for our team to give continuity to the Santo Domingo Tonaltepec workshop due to our lack of resources. Our team's lack of means made it impossible to collect the artifacts from the second burning and to make a joint reflection with the artisans on the outcome of the workshops. However, Kythzia and Diego worked with Colectivo 1050 to continue developing new projects in the Santo Domingo Tonaltepec community to promote traditional pottery techniques. And supported me as liaisons to continue the dialogue and resolve any conflicts with the artisans up today.

The San Pablo Tijaltepec project could have been continued for a longer time and a final artifact materialized for the commercialization phase. It would have marked a turning point in the project. Various topics could be followed up on with the facilitators through the workshops: elaborations, the process, the finishes, the materiality, time management, the artisans' concerns. Although the entire team could not continue the workshops due to limited resources, the facilitators continued their reflections and the artisans continue their market initiatives to this day. Ana Paula, Laura and Maddalena discussed how to continue after the San Pablo Tijaltepec artisan group's (Ñaa Ñanga) successful participation in the Expo-venta craft fair. Ana, Laura, and Maddalena proposed that Ñaa Ñanga expand their training to other women in the community in order for them to lead the efforts to transmit their experiences in the workshops within the community. This proposal generated disagreements among the artisans. Due to the lack of resources and the additional effort it entailed for the team, the facilitators are forced to postpone the initiative.

While this project's practice has concluded, its research process has opened many other parallel avenues for collaborators. We should study the impact of workshops on the community at the social and economic levels. The CADA foundation continues to seek financing and return to México to offer new workshops and give continuity to the projects started under this initiative. The anthropological opinion on the San Pablo Tijaltepec blouse, which was promoted by a team of collaborators, is another point that opened debates on indigenous peoples' right to identity as it is represented in their designs and other cultural expressions, as well as about pre-Hispanic techniques. The Colectivo 1050 and facilitator team also talked about organizing a round table event between the two communities that requested an anthropological opinion, as well as a debate in Oaxaca to continue discussing the issue of misappropriated crafts.

This project has also opened avenues for future research that has yet to be organized. For example, the social design practice I describe in this work has had an impact on the social and economic aspects of the community. These changes are important to study in detail so as to assess what sorts of lasting consequences can come from the practice. Another potential for future research is to "measure" these impacts in the artisan communities. It is necessary that we open the debate about methodologies of "how" and "what" to measure for future social design practices. In this sense, I believe that some of the key questions for future work in this field are aimed to consider how to best facilitate meaningful engagement and correspondence. This should not be another attempt to turn design into another "specialty science" or revive the humanitarian Western discourse. Instead, future work must be done that considers the fundamental components of social design, how it should be performed, who it should serve, and what its consequences can entail.

Evidence regarding the effectiveness of our approach in this work is still limited and is difficult to assess, but this research suggests that substantial progress can be made if social design practice were centered on a common agenda to initiate collective impact. As I explain in Chapter 4, I sought to encourage the participation of all. During our collective discussions about cultural heritage and the community, it became clear to me that none of the community members were fully aware of the possibly dangerous scenarios that could result from intrusive and non-ethical approaches. We are all affected by larger structures of social life—including identity and the role of national and foreign policies that shape our identities. Social problems in our communities do not arise simply from the social realm, but from an interconnection of the social, cultural, and economic. Our practice as social designers should begin to connect those aspects of all our lived experiences in order to paint a clearer picture and more directly address the issues that we face together. Unfortunately, these sorts of projects that aim to fix some social issue tend to focus on independent actions, and they are not often brought together by a common agenda.

This is not because such an effort is impossible, but rather because it is rarely attempted (Kania and Kramer, 2011).

In a voracious capitalist economy where consumerism has set the standards of well-being, it is necessary to vindicate a reflection of our social processes and ourselves. In a tradition of making, and without the need to be associated with the past, this process is a transformative one. It adds another layer of time and measurement, and allows our critical reflection and a dialogue within ourselves and with the world surrounding us. For future generations, we are leaving behind a multitude of societies based on immediacy with no meaning for memory. We should all consider it essential to foster critical and analytical visions of our surrounding world so that new generations can assume roles aimed more truly toward making progress. In the act of making, individual attributes and agencies can be promoted, communities can be strengthened, and positive changes can materialize based on a mutual correspondence.

This research shares how a social design process can lead to practices of correspondence and a form of exchange that expands market opportunities. Ñaa Ñanga, the collective founded by the artisans of San Pablo Tijaltepec, has collaborated with the CADA Foundation for four years to revive traditional techniques and improve access to tools. The collaboration has aimed to reinforce the artisans' cultural heritage through making. Today, thanks to the initiative, the collective has consolidated and acts as an independent organization to foster new collaborative work structures and business development.[1] A further collaboration was established to assist the collective with a communications strategy where the artisans chose a figure for their logo, developed a brand strategy, and cemented a digital presence for marketing. They obtained a venue on Facebook and Instagram, and generate and maintain their group profiles for themselves. The collective was also provided with a mobile phone through the collaboration so that they could communicate and establish contacts for themselves. The use of technology, greater access to global information, and communication capacity and autonomy that new technology can provide is a consequence which has yet to be evaluated.

The San Pablo Tijaltepec artisan women's participation at the Expo-ventas generated many expectations from the textile museum in Oaxaca and in the congress. The artisans were initially weary to attend the event, but we facilitators worked with them to provide total transparency about the event and the pricing strategy for their products. We together intentionally agreed increased the price to provoke a discussion about the costs of manual labor, and then facilitators created a price breakdown based on the materials, the artisans' fees, and the museum's percentage. After the Second Mesoamerican Textile Congress (TEXTIM II) and Expo-Sale at Oaxaca de Juárez, and despite the fact that the artisans were initially reluctant to attend, they realized the importance of their craft to generate income. The Textile Museum had also commissioned a blouse order from the artisans with high-quality materials for exhibition in the museum's permanent collection. This demonstrated the value of the artisan's work and knowledge as well as their community's cultural legacy as it is represented in their craft technique.

The impact of the new economic contributions made by women within their community shifted social roles in the community. During the Second Mesoamerican Textile Congress (TEXTIM II) and Expo-Sale at Oaxaca de Juárez event, men had to remain in charge of household and care for the children. This created an unprecedented social situation in the community. The event and its consequences, including on social roles and the willingness of the artisan women to adopt new technology in their sales strategy, are important points that indicate the power of a social design technique and its ability to foster connections between the local and the global. Both on a personal level and for the team, I also consider the issuance of the anthropological opinion on the San Pablo Tijaltepec blouse as important impact. The opinion carried implications for the San Pablo Tijaltepec community, and for all indigenous communities, regarding their right to identify through their designs and in their cultural expressions. The opinion also had a deep impact on the research team and led to many questions about our participation, the difficulties we dealt with in acting without intrusion, our positions as reflexive facilitators, and our extended relationships with the artisan communities.

After the artisans' success at the event, an Oaxacan commercial brand showed interest in their artifacts and approached the Ñaa Ñanga artisans about a collaboration. On the one hand, some facilitators and members of the communities felt that the proposal seemed to be an honest support for the women. On the other hand, some disagreed and felt that the brand would decontextualize the embroidery work and use their designs in other garments. The potential risk could result in the brand's theft of the community's cultural identity for monetary profit. In the hearts of the facilitators, there was a desire for the community members to decide on their own future path toward protecting their artistic heritage and the iconography of their blouses. Their embroideries, however served as part of their collective identity and a decision to collaborate with the company could have resulted similarly to the case in Tlahuitoltepec and the Isabel Marant brand's manipulation of their traditional designs. As a consequence of the debate between perspectives about the partnership, a letter was drawn up on behalf of the community that demanded protection from the CADA Foundation and the Alfredo Harp Helú Foundation in order to defend the identity of the indigenous communities.[2]

I was surprised by the political turn that the issue began to take as was assigned the positions of referee and watchdog. This event seemed to move me further away from my research and my desire to maintain a neutral position. Both the commercial brand and the communities asked me to intervene on their behalf, which led to a lot of tension and disappointment. My position was and remains neutral without providing judgment on whether the collaboration was honest or not. This attempt at neutrality generated differences between the facilitators when elected not to intervene on either behalf. The facilitators held a discussion on the right of the communities to their cultural and artisan heritage in design, and we agreed to support the communities in their doing but I could not give an opinion or take sides with either the company or the communities.

Finally, the opinion was requested by the authorities. Anthropologist María del Carmen—who also wrote the protection opinion related to the case of Isabel Marant and who acted as a collaborator and advisor to CADA—drafted the opinion in June, 2018. The opinion states:

> Aunque hace algunos años la confección de esta blusa respondía al uso local, hoy día, existe un mercado fuera del municipio que busca la compra y/o venta de estas creaciones. Las mujeres de San Pablo Tijaltepec, de manera individual o grupal (no refiere a una organización especial o comunitaria exclusiva) han optado por distintas opciones para responder a estas demandas de carácter artesanal que forman parte de sus ingresos actuales. Esto, como parte de las dinámicas comunitarias actuales, ha causado algunos desacuerdos, conflictos y replanteamientos en cuanto a la enajenación de su patrimonio cultural.
>
> Although a few years ago the making of this blouse responded to local use, today, there is a market outside the municipality that seeks to buy and / or sell these creations. The women of San Pablo Tijaltepec, individually or in groups (does not refer to a special or exclusive community organization) have chosen different options to respond to these artisanal demands that are part of their current income. This, as part of the current community dynamics, has caused some disagreements, conflicts and rethinking regarding the alienation of their cultural heritage.

The long-term objective of this work is to support the artisans in creating a sustainable niche in the global market as per their requests. Success would involve a reliable income for the artisans along with a broader recognition of their history, culture, and skills that is embodied by their work. Whether the artisans would lead this engagement through the co-operative business model, which is popular and prevalent in Oaxaca, or through individual means is yet to be determined. Indeed, success could be measured by creating a sustainable model that could be adapted across regions and where indigenous communities express their desire to maintain traditional modes of living and making while also retaining ownership over their designs and histories. In any case, the continuance of elaborate artisanal craft processes must remind us not only of where we come from but also of where we could go. We all have the potential for social and economic transformation when we elicit cooperation between various knowledge forms and worldviews across time and space.

From the research, I believe that an interesting question remains about the designer's own role as a facilitator. The team and I conceived of the designer as facilitator so as to remove ourselves from the Western educational center that prioritizes particular expertise and knowledge. We believed that the designer must adopt a role in which their mission is to be a link between artisans—their craft, their traditions, their culture, their materials—and the global market in which their artifacts can serve as "products." Contemporary design and manufacturing processes are generally isolated from one another around the globe. The designs are created by and for Westerners, but they are manufactured by non-Western peoples in non-Western countries. In this project's case, the design and production processes are unified; it is the craftsmen themselves who assume both functions—design and production—and therefore control the whole process so as to modify it to meet their needs. It would be useful to explore how other communities might respond to a similar approach and to find more suitable methods to measure transformation as a major driver of change toward more ethical and democratic ways of living and making.

It is important to note that I only found a few relevant examples that described similar projects in sufficient detail to inform my own research. As an educator for many years, I find that many design schools use the concepts of "design," "social design," or "social design practice" as an ideology to justify another means of selling products within the established markets. These programs grow as a strategy to position themselves and with the intention of selling "design" as another source of profitable income. The lack of documented case studies in the area of social design practice is sadly surprising. On a more positive note, a parallel development in design for social practice and design for education is slowly growing mainly outside of the walls of the design schools.

I consider this research to provide a useful case study for practitioners, educators, and social design researchers to open new avenues and form a crossroads between fields interested in collaborative work, social design practice, and material culture. For those adopting a practical approach while working creatively with communities, conflicts can arise not only as a result of individual action but also because of misunderstandings connected to our cultural assumptions. Thus, it is necessary that we endeavor to make ethical decisions and encourage work that frames our surrounding realities in a new way that challenges expectations and what we all-too-often take for granted. While no one can say exactly what our role may be in the future, I would like to challenge social design practitioners to continue finding ways to initiate this shift in our practice.

As I have previously mentioned, I encountered a challenge funding this project as an individual after the first workshops. I realized it would be wise to structure an entity with legal and financial backing to further develop such research concerned with design and community identities. As such, the CADA Foundation was born. CADA Foundation is a platform that promotes and disseminates the research carried out in the context of this work as well as other projects developed in indigenous communities. It operates to connect institutions, communities, and individuals who are interested in providing financial or resource support for the dissemination or exhibition of academic collaborations. The mission of this platform is to share the necessary information in order to establish a sustainable model from the environmental, social, and economic point of view based on dialogue with different communities' artisanal approach. CADA contributes to design strategies during the production and distribution chain. The goal is to foster and facilitate opportunities for local artisans and suppliers as they aim to become part of a broader design network while helping to reinforce their own communities and identities.

Finally, I feel that is important to consider how craft has remained historically encompassed in a system of exchange that designers must begin to challenge. With practicing design comes a responsibility to develop new ways of thinking about design that also highlight the relationships and the reciprocal aspects of a design practice. I believe that the view of design as a social practice of correspondence, and which requires trust, is necessary for creating a more equitable environment for craft, "making," in our global context of exchange. During workshops, I define different situations as "proximities of design" among people that are related to time and space or to the process of making together. These experiences transform in order to create a framed environment containing relationships between people, places, and materials. Social design practice should be based on democratic, qualitative, communal exchange and decision-making.

When all voices partake in the decision-making process that surround making and exchange, design can be seen as a platform for decolonizing exchange relationships. I often notice a lack of non-Western voices in the international design discourse that I am familiar with. I believe that it is necessary that we begin to represent equality in the decision makings and opinions about design. As designers, we must make a conscious effort to avoid imposing dominant and subjective worldviews; we must not arrive in places where we have not been invited without considering the goals and needs of those effected by our very presence.

I have shared how individual and community attributes and agencies can be promoted through design. Communities can build their strength and resiliency based on mutual correspondence of a social design practice. This research opens new avenues for triggering cooperation and creative transformations in artisan communities. At the least, it is my hope that this work inspires a discussion about the future applications of social design practice and workshops framed through "proximities of design." This is to say that I think of this work not as an end, but as representing the beginning of new possibilities in the fields of design and practice.

Notes

1 Self-description of Colectivo Ñaa Ñanga de San Pablo Tijaltepec, at their Facebook Information page: https://es-la.facebook.com/pg/blusastijaltepec/about/?ref=page_internal.

2 The Alfredo Harp Helú Foundation promotes actions for the good of the people of México with programs in the fields of education, culture, and sports. It stands out for

providing support to artisan workshops including training, infrastructure improvement, exhibitions, publications, and incentives for production. www.mio.org.mx/fahho/.

References

Kania, J. and Kramer, M. 2011. Collective impact. Stanford, California: Stanford Social Innovation Review.

Malinowski, B. and De La Fuente, J. 1957. La Economía de un Sistema de Mercados en México. In: Acta Anthro

pologica, Época 2, 1(2). México D.F.: Escuela Nacional de Antropología e Historia, Sociedad deAlumnos. (Drucker-Brown, S. 1982. Malinowski in México: the economics of a Mexican market system. London, Boston, Melbourne and Henley: Routledge & Kegan Paul.

Turok, M. 1988. Cómo acercarse a la artesanía. México D.F.: Editorial Plaza y Valdés. UNESCO Convention for the Safeguarding of the Intangible Cultural Heritage (ICH) 2003, Global https://unevoc.unesco.org/home/TVETipedia+Glossary/lang=en/filt=all/id=723

Afterword

Proximity and the Ethics of Engagement

Barbara Adams

As designers increasingly turn their efforts to altering conditions for those who are oppressed and made vulnerable by the systems that shape our world, stubborn questions arise around the ethics of engagement. Socially engaged projects seek meaningful change, yet often discourage dissent, reify privilege, remain agnostic about outcomes, and do little to alter larger, structural inequalities. Designers can easily exit projects deemed failures and write these off as learning experiences. Armed with empathy and expertise, but with little local knowledge, design practitioners often struggle to form equitable relationships with partners and collaborators. Moreover, most design-based efforts fail to decenter-dominant cosmologies to make room for a diversity of worldviews and ways of living in the world.

Designers have only recently taken up the issues, inequities, ineffectualness, injury related to and resulting from practices generally described as community engagement, social impact, and the like, asking how they might actively dismantle oppressive structures and systems. Notably, Carmen conducted her ethnographic and design work in Oaxaca during an unprecedented moment when practitioners began to seriously question design's oppressive history and harmful practices. As an ethnographer, Carmen was aware of the ways anthropologists have been engaged in this sort of self-critique for decades (if not longer). The reflexive turn of the mid-1980s prompted anthropologists to acknowledge their culpability in colonial projects, challenging the view that the ethnographer can objectively and unbiasedly study and equitably engage people from cultures different from their own. This interrogation of the discipline produced radical ontological, axiological, and epistemological shifts, marked by a growing awareness of the ways in which all social relationships are marked by power dynamics and of their accountability to their collaborators. Ethnographers continue to grapple with these issues while designers have only recently started to seriously address how structural and longstanding inequities both shape the knowledge produced about people and their worlds and how this guides design activities and interventions.

Many emergent paradigms attest to the ways the reflexive turn has entered design practices and processes. For example, pluriversal design[1] elaborates the

Zapatista slogan, "Queremos un mundo donde quepan muchos mundos." In a world where many worlds fit, communities themselves lead design processes and control their own life projects, activating a "radical design imagination."[2] In addressing social and ecological crises, pluriversal design underscores both interdependence and autonomy in world- and future-making. Similarly, advocates of transition design[3] urge practitioners to recognize the interconnectedness and interdependency of social, economic, political, and natural systems, calling for "cosmopolitan localism," a site-specific approach where design activities are developed and tailored for local social and environmental conditions. Transition design aims for deep change, challenging existing paradigms, envisioning new ones, and demanding radical and equitable social and environmental transformation. Demands to decolonize design gained force in 2016[4] when a group of emerging design scholars released their manifesto calling for decolonization—as a practice—to orient all design activities to advance "ecological, social, and technological conditions where multiple worlds and knowledges, involving both humans and nonhumans, can flourish in mutually enhancing ways."[5] More recently, discourse and practice have focused on steps toward design that is just, equitable, and collaborative, calling on creative practices to address the deepest challenges marginalized communities face.[6] Design justice, as a framework asks us to consider and act on the ways design distributes benefits and burdens and how the "matrix of domination" (white supremacy, heteropatriarchy, capitalism, ableism, settler colonialism, and other forms of structural inequality) plays a role in this.[7] Justice-based design practice includes communities in meaningful participation in design decisions and recognizes and values community-based, indigenous, and diasporic design traditions, knowledge, and practices.[8].

This rough and incomplete recap provides just a brief sketch of recent issues framing the work of social designers today. This work is haunted by the legacy of white supremacy and framed by difficult questions and complex challenges. How can design practitioners work with people to develop new ethical orientations? (How) can social designers build alliances, solidarity, and equitable relationships with their collaborators—particularly when these relations are embedded in unjust systems? How might designers phase out the need for social design, making small yet significant changes in people's lives while also working to transform broader overarching systems and structures? These questions are not rhetorical. They call for action-based design responses—a daunting task, given that even the notion of justice itself is oriented around Western conceptions of how we can best live together in (and with) the world.

In response to these pressing questions, Carmen offers an approach she calls "proximities of design." This strategy acknowledges and develops the ways in which designers create situations and cultivate conditions where people can creatively build connections with other people, with place, with materials, and with individual and collective practices. This method, although functioning in a framed and designed environment, functions as a flexible system that adapts and shifts through participation and collaboration. Carmen understands proximities of design as

> a way to approach and contextualize in order to bring about dialogue among all participants, not only in a verbal way, but also through gestures, movements, situations and physical distance among people, or in the way that material is used, and artefacts are made.[9]

This capaciousness in terms of form, she argues, has the capacity to forge new understandings of identity and to foster practices of collaboration in creatively negotiating cultural differences.

Following Hannah Arendt (1958),[10] who argues that the designer, artist, and craftsperson, as homo faber, create the things (material objects, stories, and other types of artifacts) that facilitate meaningful action, Carmen stresses the importance of making. In her work, she experiments with how processes of making can transform social, political, and economic situations, altering people's life chances. Arendt (1958) argues that human artifice (fabrication by human hands) provides a medium around which people can gather to create a world together. This notion, that through the process of making people can form connections and build worlds, is central to Carmen's collaborative work. Moreover, in being attuned to the embodied and performative aspects of making, Carmen recognizes how "voice" and expression are conveyed via human collective labor and in artifacts themselves. This experiential aspect poses provocative considerations for those working in participatory and social design fields. When, earlier this year, those who participated in the project were asked to reflect on the process, the statements were positive, yet not necessarily pithy. This points to a challenge for designers in terms of how to assess project-based work, even in those situations where the relationship is sustained over many years. Collaborators note significant improvements in their lives as a result of the project and the ongoing work of the CADA Foundation. They identify a better overall quality of life due to increased mobility, a sense of support, and a stable, living income. Their feedback is brief, leaving us to wonder how social designers might better invite critique from collaborators, holding design professionals accountable and responsible for the impacts (or lack thereof) of their work. Are there other tools, perhaps built into collaborative making processes themselves, that might facilitate evaluation and critique? What is outside our current perceptual orbit and how might this be expanded through collaboration with people from worlds that might be very different from our own?

The notion of "proximity," offers generative responses to these questions in terms of how we might catalyze creative forms of collaboration and critique. Carmen's use of "proximity" resonates with Deleuze's theorization of "the zone of proximity."[11] Here, there is an encounter with the other in a zone of creation where what emerges is not the possession of those in the encounter. Rather, creation is the shared event of becoming that is generated by proximity. In proximity, he asserts, we no longer occupy a stable identity but are folded into movement and a position that is nomadic. This sounds like an interesting place for social designers and their collaborators to meet. In proximity, ossified identities and ways of acting are destabilized and give way to a nomadic approach with more freedom to take experimental and exploratory detours. Framed in this way, we might see "proximities of design" as an aspirational disposition where we are constantly striving to create situations and conditions for making other things sensible, creating new alliances and forms of action.

It's been said that every new medium and technique is indicative of new social relationships.[12] If social design is to operate according to principles of justice and with a commitment to social, political, and economic transformation, the field needs new mediums and techniques. As social designers are increasingly driven to address social and political injustice, new methods and protocols are needed. This will involve taking stock of privilege, authority, and access and dispensing with deficit models. Frameworks that center damage or resilience function as failed theories of change.[13] These approaches ask individuals who are already exploited to bear the burden of change and preserve the conditions that create suffering in the first place. They position "communities" (a placeholder for those "in need" where oppression singularly defines people) as reliant on expert practitioners who function as benevolent helpers rather than dedicated allies working in solidarity to change institutional and structural inequalities. In shifting to practices of proximity as Carmen proposes, perhaps we can co-create conditions for participation that nurture relationships based in solidarity. In taking guidance from communities themselves, social designers can maintain a capacity to act, while being responsible for the consequences of acting. As Carmen notes, this is not a short-term endeavor, but an ongoing project that can move only at the speed of relationships themselves.

Notes

1 Escobar, Arturo (2017) Designs for the Pluriverse: Radical Interdependence, Autonomy, and the Making of Worlds. Duke.
2 Escobar, Designs for the Pluriverse.
3 Irwin, Terry (2015) "Transition Design: A Proposal for a New Area of Design Practice, Study, and Research," *Design and Culture*, 7(2): 229–246.
4 See: Decolonizing Design's "Editorial Statement" on their website, June 27, 2016, www.decolonisingdesign.com/statements/2016/editorial/, and their roundtable discussion, Tristan Schultz, Danah Abdulla, Ahmed Ansari, Ece Canlı, Mahmoud Keshavarz, Matthew Kiem, Luiza Prado de O. Martins & Pedro J.S. Vieira de Oliveira (2018) "What Is at Stake with Decolonizing Design?" *Design and Culture*, 10(1): 81–101.
5 Decolonizing Design, "Editorial Statement."
6 See for example the work of the Design Justice Network (https://designjustice.org/) and Sasha Costanza-Chock's book from 2020, Design Justice: Community-Led Practices to Build the Worlds We Need, from MIT.
7 Costanza-Chock, Design Justice.
8 Costanza-Chock, Design Justice.

9 See "Proximities of Design" in Chapters 1 (pp. 22–23) and two (section 2.3).
10 Arendt, Hannah (1958) *The Human Condition*. University of Chicago.
11 Deleuze, Gilles (1997) "Literature and Life," *Critical Inquiry* 23(2): 225–230.
12 Williams, Raymond (1977) *Marxism and Literature*. New York: Oxford.
13 See: Tuck, Eve (2009) "Suspending Damage: A Letter to Communities." *Harvard Educational Review* 79(3): 409–427.

Reference

Arendt, H. 1958. *The human condition*. Chicago: The University of Chicago Press.

Index

Note: Page numbers in *italics* represent figures.

Abendroth, L. M. 30
Acha, J. 52, 57, 62, 63, 97
Adams, B. 19, 95
Adams, Jonas 86
aesthetics and design practices 50–53
Aguirre Beltrán, G. 45, 53, 122
Ahmedabad Declaration of 1979 14, 39, 180
Anderson, B. 43
anthropology in Post-revolution México 44, 46–47
anti-indigenist critique 53–55
approximation 7, 19, 36, 78, 91, 142, 149, 157
Arendt, H. 192
Arroz con Leche 84
artisanal production and initiated market exposure 105; in 2018 173; on March 4 and 5 2016 168; on March 7 and 8, 2016 168; on March 17 and 18, 2017 171–173; on November 18 and 19, 2015 166, 168; in October 2015 166; in October 2016 171; on October 25, 2015 165–166; between October 2015 and March 2017 165; processes of 140
Assmann, J. 56, 96
Atkinson, Hugh 86
authenticity, notion of 26

Balaram, S. 13, 94, 95
Barrera, Kythzia 85, 140, 142, 143, 145, 149
Barro y fuego; El arte de la alfarería en Oaxaca 85
Batalla, Bonfil 54, 115, 123
Bautista, Juanita *34*, 102, 136, 138, 142, 152
Bautista, Rosalía 15, 16, 37, *98*, 99, 142, 152, 155, 158, 160, 161, 162, *162*, 166, 168, 171, 173, 181
Bell, B. 30
Bennett, J. 64, 65
Bhabha, H. K. 19, 23, 25, 26, 44
Boas, F. 46, 59
Boletín de Secretaria de Educación Pública (Bulletin of the Secretary of Public Education) 51
Bonfil, G. 43, 46, 52, 54, 56
Bonsiepe, G. 25, 27, 38, 61
Brandt, E. 11
Bratteteig, T. 105

CADA Foundation® 13, 16, 19, 78, 130–131, 168, 171, 183–188, 192
capitalism 65
capitalist well-being, concept of 61
Castillo, María del Carmen xii, 40n2, 86
CECATI 166, 32
Chapela, Rosa 87
Chatterjee, Ashoke 14, 22
Chávez, Eric 10, 86
Clancini, Néstor García 25, 55
Clarke, A. J. 62
Colectivo 1050 85, 88, 140–141, 143–144, 148–149, 168, 171, 182–183
collaboration 127; between anthropologists and social designers 27; between artisans and design practitioners 5, 9, 16, 81; between artisans and facilitators 103, 105; communications strategy 184; creative forms of 192; future work 182; with local artisans and communities 6; between participants 26; during project workshops 72; in workshops 5
collaborative design practice 5, 81, 84
collective work, notion of 47, 80
community, concept of 25, 48, 53

connected communities: culture of transience 37; encapsulation 38; group's communications 36, 37; indigenous communities 36, 38; making together 37; network of social connections 38; workshop at San Pablo Tijaltepec 37
Connerton, P. 65, 100
Cooperation 17, 24, 50, 73, 81, 186, 188
Coperative Taller Arte Papel Vista Hermosa 12
correspondence, practice of 7, 14, 29, 82, 184; cultural correspondence 60; dynamic practice of 62; equitable correspondence 30; knowledge exchange 14; mutual correspondence 31, 34, 184, 188; participation through 7, 95; social practice of 31, 188; theories of 19
Cortés, Margarita 143, 146, 147, 160, 166
cosmovision 3, 9, 33, 38, 44
Craftsmanship training 73
Cross, Nigel 104
Cruz, Demetria 160, 166, 167
Cruz, Gregoria 143, 146, 149, *150*, *167*; reading through artifact dossier *150*
cultural artifact, defined 93
cultural differences: aesthetics and design practices 50–53; anti-indigenist and post-indigenist critique 53–55; design before design 60–65; expressions of exchange 59–60; humanistic approach 65–66; imagined communities 43–44; indigenism and anthropology in Post-revolution México 44–50; resilience philosophy 55–59
cultural modernity 24

Day of the Death Celebration, The (Día de Muertos) 123
de Ávila, Alejandro 3–4, 90
deep México, defined 54
De eso que Llaman la Antropología Mexicana (On that Thing Called Mexican Anthropology) (Bonfil) 54
De La Fuente, J. 46, 47
democratic design environments: Ahmedabad Declaration 39; cultural differentiation 39; local traditional practices 39; proximities of design 39–40; reflections on design 38–39
Departamento de Asuntos Indigenistas (Department of Indigenous Affairs, or DAI) 50
Design as Future-Making (Yelavich) 7
design as practice of correspondence: building team of reflexive design facilitators 82–87; communities selection, through team connections and discussions 87–90; learning and transforming together 90–105; oral tradition and craft as cultural symbolism 74–82; permanent inaccessible road section *89*; picturing sounds and movement of making and unmaking 105–108; transformative practice and collective choice 108
design conceptualization 32, 61
designer-facilitator 139
Design for the Real World: Human Ecology and Social Change (Papanek) 100
design, in developing societies 14
design practitioners 13, 16, 18, 190–191; and artisans 5, 12, 73, 81, 102; experienced 144, 160; local communities' cultures 178; proximities of design 7; tools for social change 12
design research: artisanal production and initiated market exposure (Phase III) 165–173; general reflective meetings Part I (Phase II) 138–144; general reflective meetings Part II (Phase II) 157–159; prototyping and production (Phase III) 159–165; proximities of design, workshops research (Phase II) 144–157; and transforming methodology (phase II) 93–103
Díaz, Fátima 6, 10; facilitator meetings 83, 138, 142, 157, 158; oral workshops 11, 12, 132, 136; site visits 115, 116, 117, 119, 120
DIF (National System for the Integral Development of the Family) 16, 137
documenting processes and outcomes, meta-research 114–131; exploratory research travel and visits (2013–2014) 115–131; proximities of design 114; trips and meetings *114*
donating partners 13

El Camino de los Altos 82, 84, 109n4, 125, 173
Encuentro Indigenista Interamericano (Inter- American Indigenist Summit) 50
ethnographic research 5, 73–74, 76, 91–92, 132–138, 178
Etla Paper Factory 11, 121
Ewart, I. J. 104

exchange, expressions of 59–60
exploratory research (2013–2014): interviews *see* interviews, exploratory research (2013–2014); site and field visits *see* visits (site/field), exploratory research (2013–2014); travels *see* travels, exploratory research (2013–2014)
Expo-Sale 16, 80, 91

facilitators meetings: Oaxaca City (2014, May 14 and October 28) 138–140; Oaxaca City (2014, November 2 and 3) 140–141; Oaxaca City (2014, November 5) 141–142; Oaxaca City (2014, November 20) 142–143; Oaxaca City (2015, March 14) 157–158; Oaxaca City (2015, March 17 and May 3) 158–159; San Pablo Tijaltepec (2015, May and June) 159–165; Santo Domingo Tonaltepec (2015, February 4, 7, and 15) 143–144; Santo Domingo Tonaltepec (2015, May and June) 159–165
Familia Hernández Cooper Workshop 129–130
Fernández Ledesma, G. 50
Fernández, Omar: facilitators meetings 10, 138, 140, 142, 143, 157, 158, 159; oral workshops 11, 12, 132–133, 134, 136
field notes: documenting processes and outcomes, meta-research 114–131; social design, to approach local culture 174; workshops and facilitators' meetings 131–173
flow of framed events: San Pablo Tijaltepec, Oaxaca 136–138; Santa María de Atzompa, Oaxaca 134; Santa María Yavesía, Oaxaca 135; Santo Domingo Tonaltepec 146–149; Santo Pablo Tijaltepec, Oaxaca 152–157
Fondo Nacional para el desarrollo de la Artesanía (FONART) 84
Forcella, Maddalena 6, 10, 83; design workshop 151; facilitators meetings 84, 141, 142, 143, 144; K'un K'un project 123; oral workshops 132, 134, 136, 142; reviewing new samples *166*; travel 122
"forging of a nation" concept 54
Forjando Patria (Forging a Nation) (Gamio) 46
Fuentes, Ana Paula 84, 85, 88, 125, 138, 142, 151, 152, 157–159, 161, 171, 173, 182

Gamio, M. 45–48, 50, 51, 53, 54, 65
García Canclini, N. 25, 55, 142
García, Natividad 15, 16, 142, 152, 155, 158, 160–161, *163*, *164*, 166, 168, 171
García, Sebastiana 15, 16, 142, *151*, 152, 155, 158, 160–161, *163*, *164*, 166, 168, 171–172, *172*, 180, 182
Gächter, Priscila 87
Geertz, C. 7, 19, 62, 63, 75
Gell, A. 59, 63, 64
general reflective meetings *see* facilitators meetings
Gislev Kjærsgaard, M. 11
globalism 3–4, 7, 23–27, 32, 37–38, 40, 74, 179, 181
globalization 23; backroads 61; cultures of transience 25; homogenization patterns 25; indigenous population 24; more developed Western cultures 24; nation-state, idea of 23–24; socio-cultural homogenization 38; technological 38; and technology 27; transformational processes 36; workshop facilitators 26
Gregory, J. 93
group consciousness 48
Guiteras Holmes, C. 54, 122

Hernández Sánchez, G. 64, 89, 129
How Societies Remember (Connerton) 100
humanistic approach 19, 43, 60, 65–66
Hunt, J. 8
hybrid cultures 25

IDEO 35–36
imagined communities 43–44
"India Report" 14
indigenism in Post-revolution México 45–46, 48, 50
Instituto Nacional Indigenista (National Indigenist Institute, or INI) 48–49
intangible artifacts 18, 75, 87–88, 91
International Council of Societies of Industrial Design (ICSID) 14
intersubjectivity, concept of 44
interviews, exploratory research (2013–2014): with Ambar Past (2014, February) 121–122; to Ana Pellicer and Luis Morales (2014, November 15) 130; with Aurelia Zebadua, Luis Morales, and Ana Pellicer (2014, November 8) 124; with Carmen Pinoco and Director Miguel Ángel Alcantar García (2014, November 14) 129; with Luis Morales (2014, February) 122–123; to Marina Fraga (2014, November 13) 128

justice, notion of 92, 191

knowing by design, concept of 104
knowledge pieces 11
Knowles, C. 58, 61
Kopytoff, I. 60, 64
Krippendorff, K. 10
Kuchler, S. 98

Lacapra, D. 107
La Casa del Artesano 127
La Fábrica de Papel de Etla 12
Latour, B. 64
learning and transforming together: artisans' requests for support 91; craftsmanship and training 91; for cultural transmission 91–93; design research and transforming methodology (phase II) 93–103; forms of making 93; initial ethnographic research and oral workshops (phase I) 91–93; material of oral tradition 91–92; project methodology, development of 90–91; prototyping, artisanal production and market exposure (phase III) 103–105
local community: adoption 24; artisans to participate in workshops 80; authenticity, notion of 26; democratization of communication 26; and design practitioners 29; experience and deep relationships 83; "in- between" spaces 25, 26; with reflexivity 16; superior culture 24–25
Los artesanos del porvenir (*Artisans of the Future* 1995) (Pellicer) 56
Los Peligros de las Almas (The Dangers of the Souls) (Holmes) 54

"*maestros artesanos*", 6
making and unmaking process 7; acculturation process 97; artifact's production 98; artisans and facilitators 95–97, *97*, *98*; collaborative process 101, 104; crafts, practices of 97–98; cultural artifacts 99; design research and transforming methodology (phase II) 93–103; environment-behaviors *102*, 102–103; experimenting together with new materials artisans *100*; group reflection 101; materials and techniques 96; movement of 105–108; oral workshops 93–94; organizational routines during design workshops *99*; participation 102; participation through correspondence 95; proximities of design 95; workshops and post-reflections 96; workshop transformations 102
making things together, process of 15, 17, 19, 28, 32–36, 108
making together, action of: artisan communities 33; craft technique and material exchange 33; relationships through and within space *33*; social design practice 27–28, *30*; social interactions 33
Makovicky, N. 61
Maldonado, T. 27
Malinowski, B. 8, 47, 48, 57, 59, 60, 181
Malvar, Carmen: design workshop 145, 149, 151; facilitator meeting 140, 141, 142, 143, 149, 157, 158; field observation *133*; map tracing travel *124*; off-center position in relation to communities *140*; oral workshop 132, 134, 136; proximities of design 191–193; reading through artifact dossier *150*
manual virtuosity 62
Margolin, V. 88
market, concept of 156
market exposure 104–105
Martín Juez, F. 64
Martin, Rut 86, 138, 142, 173
meetings *114*; *see also* facilitators meetings
Metcalf, James 32, 56, 84, 122, 124, 126–129, *128*, *129*, 175
Métodos de Dibujo (Drawing Methods) (Maugard) 51
Mexican revolution 43
Mexico Distrito Federal 6
México Íntegro (Sáenz) 47
México Profundo (Deep México) (Bonfil) 54
Michoacán 6
Mier y Terán, Diego: design workshop 145; facilitator meetings 138, 142, 143, 149, 157, 158; member of Colectivo 1050 artisan cooperative 85–86, 148, 182; Utopia workshop 85; workshops *161*, 162, *169*
Mignolo, W. 25, 26
Mindling, E. 85, 88
Mixtec region 3, *4*, *63*
modern culture, system of 62
modernism 22–23, 182
modernity, concept of 22, 24, 39, 43, 50, 66
modernization 17, 22, 23, 60
money, concept of 156
Monteforte, Guillermo 26

Morales, Ludovica 86, 134, 136, 145, 151, 155
Morales, Luis 84, 93, 122, 124, 130, 134, 136, 142, 145, 151, 155
Morris, William 32, 61
multidimensional message, defined 95
Myers, F. R. 63

Ñaa Ñanga (Mixtec for Toy Women) 16
National Institute of Design (NID), Ahmedabad, India 14
nationalism, concept of 24, 43, 47, 50
nation, concept of 53, 54
Nazaria *150*, 160, *161*, 166
Nicolás, Rosario 15, 16, *99*, 142, 152, 155, 157, 158, 160, 161, 162, 166, 168, 171, 178
Novelo, V. 57, 59, 97, 181
nuclear society 53
Nussbaum, B. 35

Oaxaca Close Field Work Final Stage 131
Oaxacan Textile Museum 80
Obregón, Alvaro 50
ofrendas 123
Ojo de Agua Comunicaciones 26
oral tradition and craft: "a-capitalist culture" 77; aim of workshop program 76; artifacts 74; artisan communities 80; Colectivo Ñaa Ñanga of San Pablo Tijaltepec 78; community traditions 74–75; cross-cultural approximation 78–79; as cultural symbolism 74–75; design problem 81; ethnography of speaking 75; facilitators and artisans 75–77; Infographic Community's core *76*; initial workshops 81; Interpretation of Cultures 75; Mixtec Highlands in rural Oaxaca 76; observation and learning 77; process of reflection and collaborative design practices 81; project workshops 75; proximities through design 78; at San Pablo Tijaltepec 78; Sistemas Normativos Indígenas system 79–80; Time and Space in a Mayan Oral Tradition 75; visiting communities *79*
oral workshops: for cultural transmission 91–92; and initial ethnographic research 91–93; material of oral tradition 92; San Pablo Tijaltepec, Oaxaca (2014, February 22) 136–138; Santa María de Atzompa, Oaxaca (2014, February 8) 132–134; Santa María Yavesía, Oaxaca (2014, February 15) 134–135; *see also individual entries*; workshops
Organizational Routine (OR) *99*, 134, 135, 136, 140, 147–148, 152, 154, 155
Ornelas, Y. 93
Outline of a Theory of Practice (Bourdieu) 63–64

Papanek, V. 27, 38, 61, 100
Patzcuaro-Morelia and Mexico City 131
Pellicer, Ana 84, 122, 124, 126, 128–130, 142
Pellicer, J. 32, 56, 122
Peña, Gómez 44
"the people of the clouds" or "the people of the rain" 3
pepenado *34*, 136
petate 154
Phenomenology of Spirit (Hegel) 64
picturing sounds 105–108; collaboration between facilitators and artisans 106; empathic unsettlement 107; and movements 106–107, *107*; samples of resulting prototypes *106*; through narration 107–108
Pimentel, Francisco 45
plagiarism 27
play-dough facial expression exercise 156, *156*
post-indigenist critique 53–55
Pozas, Ricardo 53, 122
practice-based research, defined 8
pre-capitalism, notion of 60
PROADA 94
product family, concept of 160
production, design research (Phase III): San Pablo Tijaltepec and Santo Domingo Tonaltepec (2015, May and June) 159–165
Product Symbolism of Gandhi and Its Connection with Indian Mythology (Balaram) 93–94
project development 5–8; community workshops 5; phases and facilitator team development 5–6, 8–16
ProjectH 35
prototyping 103–104; design research (Phase III) 159–165; San Pablo Tijaltepec (2015, May and June) 159–165; Santo Domingo Tonaltepec (2015, May and June) 159–165
Prototyping the Social (Hunt) 8

proximities of design 7, 18, 114; democratic design environments 39–40; design practitioners 7; design research 144–157; documenting processes and outcomes 114; making and unmaking process 95; oral tradition and craft 78; practices of 191–193; reflexive design facilitators 82; San Pablo Tijaltepec, Oaxaca (2014, February 22) 136–138; Santa María de Atzompa, Oaxaca (2014, February 8) 132–134; Santa María Yavesía, Oaxaca (2014, February 15) 134–135; Santo Domingo Tonaltepec (2015, February 11, 12, 13) 145–149; Santo Domingo Tonaltepec, Colectivo 1050 (2015, February 24) 149; Santo Domingo Tonaltepec, Oaxaca (2015, February 26 and 27) 149–150; Santo Pablo Tijaltepec, Oaxaca (2015, March 2, 3, and 4) 151–157; social 33; through making and materiality 33; through space-time 33
proximity, notion of 192

Quiroz, Laura 9, 37, 86, 120, 136, 138, 141–143, 151, 152, 160, 166, *166*

Ranjan, M. P. 94
Raza Cósmica (Cosmic Race) (Vasconcelos) 47
Recchia, F. 95
Redfield, Robert 47
reflexive design facilitators: Ana Pellicer 84; building team of 82–87; co-creative communities 87; Diego Mier y Terán 85, 86; Eric Mindling 85; facilitators' workshops 82; Fátima Díaz 83; fields of graphic and editorial design 85; Kythzia Barrera 85, 86; Laura Quiroz 86; local authorities 83; Luis Morales 84; Maddalena Forcella 83–84; metal craftsmanship 84; proximities of design methodology 82; role of craftsmanship in oral tradition 84; social design process 82; team layout, basic infography *83*; traditional techniques 84
reflexive design facilitators, building team of 82–87
reflexive practice, concept of 8
reflexivity, challenges with 16–18
resilience philosophy: artifact production 57; collective memory 56; communal developments 55–56; craftsmanship and artisan production 57; cultural artifacts 57–58; facilitators and artisans 59; forms of making 58; indigenous communities 57–59
Roa Bastos, A. 52
Rockefeller Foundation 36
Ruby, J. 106–108
Ruíz, Rufina 15, 161–162, *163*, *164*, 165–166, 172

Sáenz, M. 47, 50, 51, 122, 123
sale, concept of 156
San Agustín Etla, Oaxaca 12; old Vista Hermosa hydroelectric plant 121; site visit 120–121; Taller Arte Papel (Paper Art Factory) 120–121; visits (site/field), exploratory research (2013–2014) 120–121
San Bartolo Coyotepec, Oaxaca 9, 119; visits (site/field), exploratory research (2013–2014) 119
Sánchez, Florencio 116
Sánchez, Paula 116
San Cristóbal de Las Casas, Chiapas 6, 11, 121; interior spaces at CECATI 133, *125*; travels, exploratory research (2013–2014) 125, 126
San Juan Chamula 125
San Marcos Tlapazola, Oaxaca 8–9, 115–116
San Martin Tilcajete, Oaxaca 9, 116–117
San Pablo Tijaltepec, Oaxaca 9–13, 15, 73–74, 88, 89; exchange of tools and techniques *170*; experimenting with new materials *154*; flow of framed events 136–138, 152–157; oral workshop 136–138; proximities of design 136, 151–152; samples of resulting prototypes *170*; visits (site/field), exploratory research (2013–2014) 120
San Pedro Cajonos (Villa Alta, Oaxaca) 117–119
Santa Clara del Cobre, Michoacán 6, 126
Santa María de Atzompa, Oaxaca 88; flow of framed events 134; oral workshop 132–134; proximities of design 132–133
Santa María Yavesía, Oaxaca 12, 88; flow of framed events 135; oral workshop 134–135; proximities of design 134–135
Santo Domingo Tonaltepec, Oaxaca 13, 73–74, 89, 149–150; children clay sample *148*; Colectivo 1050 149; collecting images *150*; flow of framed events

146–149; proximities of design 145–146; workshops research 145–149, 151–157
Schon, Donald 8
Second Mesoamerican Textile Congress (TEXTIM II) 16, 80
Secretariado de Educación Pública (Public Education Department) 51
selecting communities through team connections and discussions 87–90
self-criticism, process of 139
self-objectification process 43
silkworm farming 117–119, *118*
Simonsen, J. 105
Sistemas Normativos Indígenas 125, 136
Slo'il jchiltaktik 92
Smith, Adam 22, 60
social cohesion 25
social design practice 23; action of making together 27–28, *30*; to approach local culture 174; with artisans 31; communities' permeability and resilience *28*; continuity, impact, legacy, and future for 182–188; creating both material and immaterial things 29; design practitioners and local community 29; greetings and sharing food *35*; indigenous communities 31; individual and communal characteristics 34; intrusive global factors 28; knowledge and action, relationship between 29; proposals as facilitators 29; radical proximity 35; revisiting project goals 178–182; theory-of-practice discussions 27; tools in 30; use in workshops 30
Social Exchange Through Design 16
Sofaer, J. 63
sounds, defined 106–107
Stallybrass, P. 90
Steiner, C. B. 63, 64

tabula rasa, existence of 22
Taller Arte Papel (Paper Art Factory): visits (site/field), exploratory research (2013–2014) 120–121
tapete 152
task-artifact cycle 13, 15, *106*
team meeting: at café center *87*; of facilitators and collaborators 10
Textile Museum, Oaxaca 16, 123; training workshop 16
TEXTIM II 91
theory of art 63
to San Cristóbal de las Casas, Chiapas: travels, exploratory research (2013–2014) 123–124
Toscano, S. 62
transculturality, concept of 44
transformative practice and the collective choice 108
travels, exploratory research (2013–2014): to Chiapas and Communities in Oaxaca (2014, February) 121–122; to Chiapas/San Cristóbal de las Casas (2014, February) 122–123; to San Cristóbal de las Casas (2014, November 9) 125; to San Cristóbal de las Casas, Chiapas (2014, November 6) 123–124; from San Cristóbal de las Casas, Chiapas, to Santa Clara del Cobre, Michoacán (2014, November 11) 126; to San Juan Chamula (2014, November 10) 125
trips *114*
Tunstall, E. 19, 36, 58, 94
Turok, M. 57, 59, 122, 181

United Nations Educational, Scientific and Cultural Organization (UNESCO) 88
United Nations Industrial Development Organization (UNIDO) 14
unmaking process *see* making and unmaking process

The Various Arts: Medieval Treatise on Painting, Glasswork and Metalwork (Theophilus) 126–127
Vasconcelos, José 46, 47
Villoro, L. 19, 24, 25, 45, 54, 55
visits (site/field), exploratory research (2013–2014): to Chiapas and Communities in Oaxaca (2014, February) 121–122; to Familia Hernández Cooper Workshop (2014, November 15) 129–130; Oaxaca Close Field Work Final Stage (2014, November 21) 131; to Patzcuaro-Morelia and Mexico City (2014, November 16 and 17) 131; to San Agustín Etla (Etla, Oaxaca) Taller Arte Papel (Paper Art Factory) (2013, February) 120–121; to San Bartolo Coyotepec (Oaxaca Center District) (2013, February) 119; to San Marcos Tlapazola (Tlacoula de Matamoros, Oaxaca) (2013, February) 115–116; to San Martin Tilcajete

(Ocotlan, Oaxaca) (2013, February) 116–117; to San Pablo Tijaltepec (Tlaxiaco, Oaxaca) (2013, February) 120; to San Pedro Cajonos (Villa Alta, Oaxaca) (2013, February) 117–119; to Textile Museum in Oaxaca (2014, November 1) 123

Wed Do Sed Association 117
Werbel, Emmett 86
Western market system 22
Widman, Nuria 86
workshops: between 2013 and 2018 5; artisanal production and initiated market exposure (design research) (Phase III) 165–173; community workshops 5; ideas of co-creation 5; mutual exchange of information 5; oral *see* oral workshops; participation, inclusive and democratic 14–15; prototyping and production (design research) (Phase III) 159–165; proximities of design (design research) (Phase II) 144–157

Yelavich, S. 7

Zapatism 54
Zapotecs 3
Zeisel, J. 102
Zuckermann, López 127

For Product Safety Concerns and Information please contact our EU
representative GPSR@taylorandfrancis.com
Taylor & Francis Verlag GmbH, Kaufingerstraße 24, 80331 München, Germany

www.ingramcontent.com/pod-product-compliance
Ingram Content Group UK Ltd.
Pitfield, Milton Keynes, MK11 3LW, UK
UKHW020140200926
475702UK00005B/841
9781032420134